Encountering Adamastor: South Africa's founding geographers in time and place, Volume 1

Published by SUN MeDIA Stellenbosch under the SUN PRESS imprint.

This publication was subjected to an independent double-blind peer evaluation by the Publisher.

First edition 2016

ISBN 978-1-928357-27-8
ISBN 978-1-928357-28-5 (e-book)

DOI: 10.18820/9781928357285

Set in Warnock Pro 10/13.5

Cover design and typesetting by SUN MeDIA Stellenbosch.

SUN PRESS is an imprint of AFRICAN SUN MeDIA. Academic, professional and reference works are published under this imprint in print and electronic format. This publication may be ordered directly from www.sun-e-shop.co.za.

www.africansunmedia.co.za
africansunmedia.snapplify.com (e-books)
www.sun-e-shop.co.za

Volume 1

Encountering Adamastor

South Africa's founding geographers in time and place

W.S. Barnard

SUN PRESS

Contents

Foreword

Life-histories and the biographical essay

The volume of W.S. Barnard is a first in a series of life-histories of the founding geographers in South Africa published by the Centre of Geographical Analysis at Stellenbosch University. Life-histories are reported in five ways: the *commendation* lauds the winner of an award; the *obituary* by necessity speaks well of the dead; the brief entry in a *dictionary* or encyclopedia is highly stylized and constrained by editorial guidelines; in the *autobiography* the author tells his own story in his own way; and, finally, the *biography* comes in a range of formats and structures. At the one extreme is the complete life-history, written by a specialist following the historiographical method and based on the critical assessment of primary sources; at the other is the belletristic overview: descriptive, anecdotal, facile and artful.

Few geographers have risen to the rarefied heights of formal life-histories. One or two have tried their hand at autobiography (e.g. Haggett 1990), but very few have appeared in international or even national biographical dictionaries[1] and only the exalted elite have qualified for a complete biography. One individual in the sphere of geography who achieved all four distinctions is the nineteenth-century German scientist Alexander von Humboldt (1759-1859). He is the subject of a number of biographies (e.g. Beck 1959; Kellner 1963) as well as a belletristic book (Botting 1974) and numerous entries in biographical dictionaries. Of course, Von Humboldt had an adventurous life and his influence extended far beyond the confines of geography. Within their own literature, geographers have fared better. A collection of life-histories, some admittedly on very obscure persons, appeared between 1977 and 2000 in the series *Geographers: Biobibliographical Studies*, while Ann Buttimer (1993) persuaded twelve contemporaries to talk about their careers, audio-taped the interviews and scripted them into brief reports. Both collections are useful but they are cast into rather rigid moulds.

Among the different genres of biography there should be a format that condenses the life-histories of the South African founders to manageable proportions and at the same time presents a balanced perspective on their persons, experiences, and texts. Moreover, geography must be the leitmotif explaining their lives as geographers. The form chosen here is the biographical essay or biographical sketch, a compact and flexible narrative which also uses primary sources to extract the essence of a geographer's thought and practice without following him through all the humdrum of everyday life (Engelberg & Schleier 1990).

Each essay in this book moves through the four phases of the historiographical method, commencing with general *goals* and transforming them into specific *questions*. The second step is the collection of *data*: formal and informal publications by the subject, primary manuscripts in archives, autobiographical and biographical notes, interviews with family, relations, former colleagues and students, contemporary publications to provide context, and visits to places where the subject lived and worked.

The third step is to order, collate, analyse and *interpret* the data according to what the biographer regards as important. Lytton Strachey (1916), an early exponent of the biographical essay, plumbed the psychological contradictions of public figures who defined the Victorian Age, in some cases delving deeper than evidence allowed. Similarly, the biographer of the artist often looks for hidden traits and motives, "the figure under the carpet" (Edel 1981: 17). When a geographer is the subject, an appropriate key to interpretation is place, the sites and areas which held meaning for him or her.

Lambert & Lester (2006) present life-history as a trajectory through space and time which intersects other trajectories at specific places and becomes entwined in local discourses. For present purposes two sets of places are important: the places where the subject spent his youth and acquired his moulding as a geographer and the places where he practised his geography. At each place the subject was involved in new conversations from which he drew his texts. Places and discourses form the context, and the biographer brings them as close to the surface as possible. Texts blend and tie these elements together with the subjects' sentiments, emotions, experiences, and professional judgements.

Finally, the interpretation, the meaning which the biographer extracts from the processed data, is *written up*. As historian, the writer has to do justice to the data but equally important also communicate with the readers. This implies selecting, ordering, and formatting the interpreted material and presenting it in such a way that description is in balance with interpretation.

Two matters require elaboration. The first concerns the rider, "time and place", which is part of the subtitle of this book. For many biographers the "times" of a subject grandly involve only national and global events. Local events, however, are just as important, and so is the state of science during the period under review: the prevailing paradigms, the theories they spawned, and their impact on geography. Similarly "place" implies not only the space perceived by the subject, which may be city or country or continent, but also society. Always, a geographer's texts are cast within a multidimensional mould; at the same time these texts inform the images that people conjure up of particular places.

The second matter involves the biographical essay as literature. The essay is a narrative, a way to link experiences in the life of a person who lived perhaps fifty or a hundred years earlier. The biographer stands outside these events and constructs a story with an opening, plot and closure, "controlled fiction" as Livingstone (1999: 21) calls it. Yet, the biographer cannot go further than the evidence allows and can only select, summarize, and emphasize. Thus, it is the structure of the narrative which is manipulated: the facts remain sacrosanct.

The importance of structure aside, this book approaches the founding geographers as persons, as distinctive human beings. The common interpretive scheme does not negate their individuality: none of their life-histories followed the same trajectory, their places of intersection differed, and their texts diverged in form and content. The challenge is to balance these differences against their commonalities. "Writing biography", Barnes (2001a: 1416) remarked, "is not easy. It requires a set of negotiations to be made between individuals and their context, between facts of life and its narrative form, and between subjects and their author. Writing ninety pages of equations, even of quantum mechanics, might well be an easier task."

The founders shared one crucial context which served as a common point of departure: the vast and varied southern African milieu. Chapter One develops guidelines for the series' volumes as a whole. It explores some approaches to the history of geography; selects a suitable format for the "life-histories" of the founders; identifies the individuals who qualify; and formulates a structure within which each person's story may be told. Chapter Two introduces this macro context and reviews the subcontinent's changing nature up to 1940: the contested political and social discourses, the environmental and institutional framework within which South African science developed and eventually geography's establishment as a university discipline. Chapters Three to Five, the real purpose of this book, focus on the founders (Passarge, Schwarz, Serton) as individuals: each in his own time, place and circumstances, each in response to the environmental, scientific and political issues of the day, each in his strengths and weaknesses.

Endnote

1 Professor Barnie Barnard had a huge influence on us as young geography students at Stellenbosch University in the 1990s. His enthusiasm for his beloved subject geography was inspiring and his honours module on geographical thought legendary. He constantly reminded and taught us to question the discipline's growth, key moments in history, paradigms and leading scholars' contributions to geography.
Due to Barnard's untimely death in 2010, Volume 1 on the founding geographers in South Africa was not a hundred per cent completed. Barnard was a meticulous scholar but we could not among the many files and folders trace a reference list for the book. We nevertheless attempted to find these sources, were to some extent successful, but unfortunately a large number of the documented references (especially archival material) are not listed in the reference list. The language editing of Pieter de Necker and Tiani Moolman's assistance in sourcing the missing references are greatly appreciated.

Series Editors: Ronnie Donaldson & Gustav Visser, 2016, Centre for Geographical Analysis, Stellenbosch University.

CHAPTER

1

Telling Geographers' Lives

In Luís Vaz de Camões' epic of 1572, *The Lusiads*, the future South Africa entered world literature as the lair of the Earth giant Adamastor, who threatened Vasco da Gama's mariners on their way to the East. Although Adamastor personified the Cape of Storms, "that mighty and hidden Cape... that neither Ptolemy, Pomponius, Strabo, Pliny nor any of past times ever had knowledge of" (Ferreira 1995: 36), he was also a figure from Greek mythology banished to the most distant corner of the Old World after his rebellion against the gods of Olympus. The first South African geographers, the generation who established geography as a professional field during the first half of the twentieth century were, like Adamastor, people of two worlds. From Europe they brought their own myths, histories, and professional training; in South Africa they had to face a harsh environment and a strained society, but also a world of fascination and challenge. The theme of these opening chapters is the counterpoint of these two narratives. Who were these geographers? How did they interweave their experiences of Europe and Africa? How were their observations transformed into texts of an international field of knowledge? These essays seek to answer these questions.

This chapter develops guidelines for the project as a whole. It explores some approaches to the history of geography; selects a suitable format for the "life-histories" of the founders; identifies the individuals who qualify; and formulates a structure within which each person's story may be told.

Exploring the history of geography: Discipline, discourse and individuals

The term "geography" was coined 2 200 years ago by Eratosthenes, the curator of the Museum of Alexandria, from the Greek words *gê* and *graphein*. Eratosthenes used the term for his reformed map of the world, but his successors found it ambivalent: while *gê* means "the Earth", *graphein* may either mean "to draw" or "to write". Hence, some ancient geographers, like Ptolemy, tried to draw the Earth's surface on maps: they were cartographers. Others, like Strabo, described regions and places verbally as background to their historical narratives.

The geography of the twenty-first century is equally ambivalent but its fault lines occupy different positions. On the one hand, the term geography refers to an ordered field of knowledge maintained and extended by professionals who call themselves geographers and have their work certified by other geographers, their peers. For the French geographer, Lefèbvre (1991), this is *la géographie des professeurs,* although many of its practitioners have nothing to do with universities. But Lefèbvre also recognizes *la géographie des banquiers,* the geography of bankers, which consists of functional knowledge about places collected by public and business concerns to optimize their operations. Both practices are called geography, and their factual bases overlap but the difference is crucial. Professional geography handles concepts that specialists weave into a coherent system of thought and use to interpret the Earth's surface. Popular geography is about information, autonomous facts that may be used by anyone (Barnard 2001).[2] This book is concerned with professional geography: the South African founders were professionals.

Geographers, Neil Smith remarked, "have been very bad at writing their own history." There are reasons for this. Geographers are a relatively small intellectual community and few hands are available to fossick the past. In their conceptualization of geography they have preferred to look in all directions but backwards. During the first half of the twentieth century geographers followed Immanuel Kant's admonition to restrict themselves to the present. The past, even their own past, was a foreign country occupied by their sister discipline, history. Although this restriction to the present eased by the middle of the century, geographers became enthralled by the wide horizons of grand theory: at first spatial theory during the 1960s and then, in reaction, Marxism and structuralism during the 1970s.

The 1980s, in yet another turnabout, brought rejection of all grand theory and the master texts on which they were based: on the one hand Kant, Comte and positivism, and on the other hand Marx and structuralism. New modes of interpretation, such as postmodernism, poststructuralism, and postcolonialism, claimed to be flexible, pluralistic, personal, and emancipating. Thus, they permeated the divisions between disciplines and replaced the rigid demarcation of research themes with an interpenetrating range of cultural studies. The new humanists had one characteristic in common: an insistence on the importance of history. Eventually geographers also turned their tracks backwards, discovered gripping narratives in their own past and, indeed, caught other specialists poaching in their backyard. For instance, postcolonial historians, sociologists, and literary critics were fascinated by the involvement of nineteenth geographical societies (especially the Royal Geographical Society in Britain) in exploration, colonization, settlement, empire building, and war (Heffernan 1996). At a local scale cultural studies discovered place, locality, landscape and regions, the markers of individual and collective identity. Moreover, historical geography revived strongly (Lester 2003), while the history of geography condensed into a distinct subdiscipline with its own specialists, institutions, and events (Driver 1995).

The historian of geography has a choice among three research strategies. The first is to target the discipline of geography itself: its schools of thought, subdisciplines, key publications, institutions, and, within this framework, its august personalities. Most studies of the twentieth century followed this approach and have been duly criticized for "presentism", exalting the present as the summit to which the past always aspired (Aay 1981). Textbooks are especially prone to fall into this trap, becoming "internalist accounts written by geographers themselves about other geographers for still other geographers" (Baigent 2004: 542). An author using such an approach demarcates the discipline according to the definition of the day, ignores work outside its boundaries, and presents the discipline as a closed system. The second strategy is to view geography's past (and present) as a discourse, a contested but continuous conversation among geographers who accept the same rules and, despite their differences, work toward stabilizing the meaning of concepts. Discourses are tied to themes rather than rigid definitions, extend the field of geography beyond the academic discipline, and accommodate changes in both scope and content (Livingstone 2000).

The third option shifts the focus from geography to geographers, the individuals who produce, maintain and apply geographical knowledge. Because this book concerns the South African founding geographers, that is people, it is obviously the preferred approach here and it is worthwhile giving some thought to its implications. The "life-history" of a particular person has, solid advantages. It provides different systems of thought with a common subject, the person or the individual; it braids the separate threads of history into a narrative played out during a particular time; it adjusts the reader's observations to the level where events occur; and, last but not least, introduces elements of human ideals and frailties into scientific writing (Barnes 2001a, 2001b; Possing 2001).

Even with their popularity with the average reader, life-histories struggle for credibility in the historiography of science. For natural scientists, biography is too personal, local, and time specific. Scientists aspire to universality; when they write they keep themselves in the background, their style is minimalistic, and their attitude is often distant and neutral. In contrast, postmodernism is deeply interested in individuals but abhors icons. This point is stressed by Michel Foucault (1970: xiv): "If there is one approach I do reject, it is that which gives absolute priority to the observing subject, which attributes a constituent role to an act, which places its own point of view at the origin of all historicity." Postmodernists deconstruct, fragment, and reject all forms of intellectual hegemony, be it the authoritive "grand old man", his master text or his metanarrative. Therefore, historians of geography who have taken the individual turn opt for micronarratives of multiplicity, uncovering aspects of their subjects not so obvious at the surface. Examples include the mystic side of two celebrated geographers (Matless 1991), a British colonial official as practical geographer (Myers 1998), the authors of an epoch-making paper (Barns 2001b), a failed explorer (Driver 2001), the commemoration of a well-known explorer at different times and places (Whithers 2003), a geography student's experience of place (Lorimer 2003), the intellectual forming of a later philosopher of geography (Keighren 2005) and so forth. These studies are imaginative, lively, probing and slightly irreverent, but often seem to miss the real thrust of a geographer's work.

What is needed is a life-history which will retain the liveliness, diversity, and revealing value of the postmodernist approach without unnecessarily fragmenting the historical pattern of geography. Geography, according to Livingstone (quoted in Barnard 2001: 7), "is an evolving tradition of inquiry – conversation woven around such themes as map, place, space, region, landscape, and the relations between nature and culture. The history of geography is embedded in this discourse and allows for progress within the tradition because new concepts are superior to their predecessors." This cluster of chapters, therefore, departs from the postmodernist position in two respects. It accepts, first, that throughout its history of 2 200 years geography has maintained a basic continuity. There were times of stagnation, to be sure, as during the Dark Ages the West forgot the sphericity of the Earth, while the heritage of Eratosthenes and Ptolemy was cherished and eventually passed on by the Arab geographers. There were also times when geography had a different function from today, as during the sixteenth and seventeenth centuries when geographers provided the technical expertise which made the voyages of discovery possible. But through the centuries the themes of geography, despite subtle shifts in emphasis, remained essentially the same. Second, certain individuals – like Strabo, Von Humboldt, Ratzel and Harvey – *did* have a deeper impact on the discourse of geography than others and therefore deserve more attention. Postmodernists may deconstruct them, while historians within the geography-as-discipline approach may hail them as "grand old men." But they remain distinct individuals who had their moments of both insight and confusion.

What genre of life-history will best capture their perspectives?

Identifying the founders: Who and why?

In any scientific community the search for founders begins with the history of the particular field of knowledge. For South African geography such a historical literature does not exist. With one exception discussed later, local geographers fared badly, not only when compared with geographers elsewhere, but also in comparison with South African sociologists (Groenewald 1984), historians (Smith 1988) and anthropologists (Hammond-Tooke 1997). What is available is superficial, incomplete and unsophisticated, and it tends to be organized on just two levels. At the more mundane level, papers follow a geography-as-discipline approach: the history of the early years of an institution (Muller 1953; Jackson 1978), the changing conceptualization of a climatic feature (MacGregor 1955), the state of specific subdisciplines (e.g. Christopher 1975), a particular genre of geography writing (Barnard 1999) and more comprehensively, but also more peripherally, the history of South African maps (Lewis 1936; Liebenberg 2004).

The second level is favoured by "progressive" geographers whose programmatic overviews of the 1980s and 1990s glance dismissively back at the 1950s and 1960s and ignore everything that went before (Beavon & Rogerson 1981; Rogerson & Parnell 1987; Crush 1991; Ramutsindela 2002). In many ways, South African geography is deeply in their debt: they fought discrimination and injustice, introduced discourse analysis into geography, and extended the field to new themes and methodologies. But they were just as shackled to the North Atlantic core as their predecessors, their materialist stance blinded them to the plurality of the discipline and, in contrast to Van Onselen's Kas Maine in *The Seed is Mine* (1999), they looked right through the individual without seeing him. Postcolonial geographers of the twenty-first century, on the other hand, have had an intense interest in individuals, but as yet there have been no geographers on their horizon (Lambert & Lester 2006).

One redeeming landmark in this wasteland is Harold Wesso's wide-ranging doctoral dissertation (1992) on the institutionalization of South African geography at school and university. Based on primary sources, logically structured and fluently articulated, the thesis traced in great detail the field's struggle to become established as a discipline, a goal attained during the second decade of the twentieth century. Wesso, however, allowed his attention to wander to university syllabi, which he found saturated with environmental determinism. For him, environmental determinism was not simply an interpretive key or a social practice of a particular time and place: it was an instrument intentionally forged to justify imperialism, colonialism and racism. Although Wesso did recognize individual geographers, he was not primarily interested in their research and texts.

While formal histories of South African geography have eschewed in-depth treatment of individual geographers, many individuals have been acknowledged in brief, appreciative commendations and obituaries. Here Conradie's commendatio (1964) and eventually Davies' obituary (1996) of William Talbot (1908-1995) are as good examples as any other. Dictionary entries usually observe a more critical distance, but only one South African geographer, John Wellington (1892-1981), has made it into an international biographical dictionary of geographers (Baker 1984). A model of the biographical essay worth emulating is historian Joan Carruthers' 2003 paper on Friedrich Jeppe (1834-1898), the amateur who mapped the late nineteenth century southern African interior. Jeppe, the first postmaster general of the old South African Republic, compiled at his own cost a map of the Transvaal and upgraded it through five new editions between 1868 and 1898. Carruthers wove a narrative of the mapping of the Highveld and Bushveld around the person of Jeppe and pointed out how the increasingly finer resolution of his maps synchronized with the emerging identity of the Transvaal, eventually surviving the severe test of the South African War.

Not all early South African geographers, however valuable their work as lecturers, teachers, researchers and supervisors, qualify as founders. It is worthwhile then to define the term "founder" with greater precision. Among the five meanings the *Oxford English Dictionary* attaches to the verb "to found", the phrase "to set up or establish for the first time, especially with provision for its perpetual maintenance" is closest to the use of the term in this investigation. "Founder", or its more elaborate alternative "founding geographer", becomes by extension "one who sets up and institutes for the first time." The founding geographers, therefore, are the originators of a specifically South African discourse of geography. A discourse is conducted through texts which, in geography, may take on many forms: books, papers, reports, maps, sketches, photographs, even architecture and human-made landscapes. Here a high premium is placed on printed texts: books and papers circulate worldwide, they are immutable and may still be read centuries afterwards. The founders are persons who gave content and structure to the emerging geographical discourse through meaningful publications. No publications, no participation in the discourse, therefore no founder! A founder need not to have been a South African or even primarily a geographer. But a substantial part of a founder's work should have addressed southern African themes; displayed an awareness of the geographical problems of southern Africa; contributed to their understanding; presented the relevant parts of the work as geography; and left a heritage on which successors could build.

A reading of South African geographical literature suggests that we should look for the founders among three groups of persons: the leaders of the first geography departments who, apart from their writing, also developed curricula, courses and research programmes; local scientists from adjacent fields who strayed into the discourse of geography; and geographers from overseas who did research in the subcontinent and published benchmark monographs.

Among the first group, the earliest lecturers at the larger South African universities come to mind, people who taught only geography: James Marnoch Hutcheon (1888-1921) who was appointed to the South African School of Mines and Technology (after 1922 Witwatersrand University - WITS) in 1917; Petrus Serton (1888-1963) at Stellenbosch in 1920; Francis Plummer (1892-1970) at the Transvaal University College (after 1930 the University of Pretoria) in 1923 and William John Talbot (1908-1995) at the University of Cape Town in 1936. Hutcheon and Plummer may be eliminated. Hutcheon, admittedly, is a marginal case. While teaching at the South African College in Cape Town during the early 1900s, he tirelessly advocated a reformed geography at school and in his first year in Johannesburg, established the South African Geographical Society. Sadly, he died young in 1921 after only four years as university teacher and without an opportunity to influence the future direction of the geographical discourse. His place was taken by his successor, John Harold Wellington (1892-1981), who led the Wits department from 1923 to 1957 and eventually built up the best publication record among the early geographers. By contrast, the statistically-minded Plummer was involved in one important report on rainfall variability and sustained the theme in three more papers, a range too specialized to qualify him as a founder.

Among the scientists from adjacent fields, the geologists Alexander Logie du Toit (1878-1945) and Ernest Hubert Lewis Schwarz (1873-1928) are the most likely candidates. Both ventured their hand at geography textbooks and served terms as president of the South African Geographical Society. However, Du Toit does not fit: he was the epitome of the field geologist, a scientist of international renown and an undaunted advocate of "drifting continents", but his perception of geography had an obvious geomorphology bias and he was never a university teacher. Schwarz lectured in geology and geography at Rhodes University College in Grahamstown and raised the environmental mirage, later known as the Schwarz Scheme, while writing a textbook, *A South African Geography* (1921). He was also the first South African Earth scientist who took his ideas into the public arena.

Andrew John Herbertson (1865-1915) is credited as the first professional geographer from overseas to visit South Africa (Wesso 1992) but the event which attracted him, the joint conference of the South African Society for the Advancement of Science and the British Association for the Advancement of Science in 1905, was also attended by other geographers who actually published papers on their experiences. More comprehensive but less known are the excellent research monographs by Karl Dove (1863-1922) on German South West Africa and Siegfried Passarge (1863-1958) on the Kalahari, both based on field research during the 1890s. The reason for their neglect is obvious: they wrote in German. The two pathfinders were followed by a succession of young researchers, including Leonhard Schultze (1872-1955), Fritz Jaeger (1881-1966) and Leo Waibel (1871-1951), all dedicated field workers who concentrated on German South West Africa and produced monographs of a standard unequalled by anything that appeared in the Union of South Africa until the 1950s. Of these five German geographers, this study focuses on Passarge: he had an enduring impact on the study of southern African stratigraphy, geomorphology and anthropology, and was the first professional to write a regional geography of the subcontinent.

From our shortlist of twelve prospective founders, five qualify: Serton, Wellington, Talbot, Schwarz and Passarge, all five white and male, and all born and educated in Europe. Each will be the subject of a biographical essay.

Closing the agenda: Demarcation and goals

How should each chapter be structured and in what larger framework should each be positioned?

The founders share two characteristics which circumscribes the project in time and place. First, all five were university teachers and their careers overlapped. Between 1905 when Schwarz was appointed at Rhodes University College and 1974 when Talbot retired from the University of Cape Town, a period of nearly three quarters of a century, at least one founder, in most years more, occupied a university chair in South Africa. Taking their participation in the geographical discourse as measure, the dates of the first and last major publication of the group provide even more satisfactory markers. Passarge's *Die Kalahari* appeared in 1904 and Wellington's *South West Africa and its Human Issues* in 1967. Monographs of this scope have a foreground, a time of preparation, research and writing. Passarge commenced research in 1895 while Wellington's honing of his manuscript continued right up to the mid-1960s. So the years between 1895 and 1965 may be regarded as the period during which the founders either dominated the geographical discourse or produced ideas which were very influential.[3] It makes sense to sequence them in the order in which they entered the discourse: first Passarge, then Schwarz, followed by Serton, Wellington and Talbot.

A second characteristic is important. With the exception of Passarge, the founders completed their careers in South Africa, identified with the country and came to see the rest of the world from a South African vantage point. The geography of the subcontinent was also a professional interest. But where was South Africa? From the 1880s to the 1950s the term "South Africa" had a wider connotation than it has today and the attention of all five founders strayed from time to time beyond the borders of the Union of South Africa proper. Here "southern Africa" is used for the subcontinent south of the Cunene and Zambezi Rivers, while "South Africa" refers to what was between 1910 and 1960 the Union (now the Republic) of South Africa. For other polities, the official names in vogue in 1960 are used: Basutoland (Lesotho), Swaziland, Bechuanaland (Botswana), Southern Rhodesia (Zimbabwe) and South West Africa (Namibia).

With the project situated in time and place, the initial goals require greater precision. At the beginning of the chapter three goals were envisaged, namely first to assess the founders as individuals and geographers; second to record and contextualize their experience of Europe and southern Africa; and third to trace the ways in which their texts weaved themes from both areas into the discourse of geography. These goals are now rephrased as a series of specific questions arranged in three clusters, indicating the trajectories along which the life and work of each founder are interpreted.

The first cluster of questions concerns the founder as person. What were his origins, appearance, personality, social class, family circumstances and education? Why did he become a geographer and in which fields did he specialize? The second cluster explores space. Where was the founder's home region and where did he study geography? Why did he come to South Africa, where was he employed and where did he conduct his research? The third cluster of questions targets the texts authored by the founder. Where and what did he publish, in which discourses did he become involved, what themes appealed to him and how were they represented in his texts? Finally, what was the impact of his work in his own time and afterwards? These questions cannot always be fully answered, but serve as guidelines for establishing a common baseline for all five founders. Each trajectory will be interpreted and narrated as a separated case.

The essays cover persons associated with a small scientific community. Why pay attention to them? The history of a particular discipline is part of intellectual history as a whole while the transmission of particular modes of thought and expression from one world region to another – in this case from the European "core" to the African "periphery" – is a universal theme. At the same time, the project seems to run against the grain of the twenty-first century: its view is from the top downwards. A history of science, however cannot be otherwise. The founders of South African geography, like scientists everywhere, were well educated and articulate persons with a middle-class background and a spatial mobility unusual for their times. Lifestyles do not make ideas irrelevant. Today, after 60 years, the founders may be accused of supporting a white hegemony which, at the close of their careers, degenerated into apartheid. The extent to which a blanket accusation of this nature is valid is considered in a later chapter.

There are other reasons why a series of biographical essays makes sense. Biography provides insight into how a creative person develops cognitive awareness and a feeling for scientific problems. Although the South African founders included no Von Humboldts or Harveys, they were solid scientists who provided a firm foundation for their successors. Biography, moreover, recognizes the nuances of scientific discourse, of the debates among discussants and of the changing meaning of a particular concept in different times and places. It also suggests new and unexpected lines of enquiry. Why, for instance, are most South African geography departments located in their universities' Faculty of Natural Sciences while their students qualify in the social sciences? Good biography, finally, does not only elucidate a person but provides an insider's view of his or her time and place, a prism of both history and geography. Geographers have a strong feeling for place and are experienced field observers: rural and urban landscapes, environmental and social problems, events and personalities, facts and myths – all contexts within which the discourses of the day are embedded.

Endnotes

1 The five volumes of the *South African Biographical Dictionary* (Beyers 1987) illustrate this point. The number of entries totals 4 500 among which about 600 cover scientists and academics of which only three are geographers: Passarge, Schwarz and Serton.

2 Most people's perception of geography derives from the subject they followed at school but school geography is not professional geography. Professional geography concerns itself with the creation of new knowledge. School geography uses this knowledge to turn learners into informed and discriminating citizens. Still, high school geography teachers were educated as professional geographers even though they do not practice it and the association between professional and school geography has always been mutually supportive.

3 For Fuggle (1967), 1917 was the year during which South African geography was established. In that year, he contended, geography was recognized by the Union Department of Education as a university subject, the first permanent lecturer in geography was appointed and the first volume of the *South African Geographical Journal* rolled off the press. This book is concerned with geography as practised by professionals. Clearly, the years during which geographers entered into discourse over southern African themes were the fateful decade between 1895 and 1905. Even if the point of departure is geography as a discipline, the choice of 1917 is disputable. The University of the Cape of Good Hope recognized geography as a subject for BA Pass in 1913 while the Victoria College at Stellenbosch started teaching a one-year course in geography in 1914 (Wesso 1992).

CHAPTER

2

The South African Context: Politics, Science and Geography

This chapter summarizes the context – the attendant circumstances – within which all founders started their southern African careers. Four elements are selected as important and presented as narratives. The opening section focuses on the changing territorial structure of southern Africa during the first four decades of the twentieth century, with Union in 1910 as the centre point. The Union of South Africa, in turn, was the outcome of a set of interactions among the subcontinent's cultural groups. The second theme traces the political discourses subsequent to these interactions, the values and concerns of the discussants and their perceptions of each other. The third theme narrows the focus to the South African discourse of science, emerging during the nineteenth century and maturing by the 1930s. The fourth theme runs in parallel with the third, that is the minor theme of geography unravelled from the major theme of science: opening a discourse, establishing institutions for the practice of geography, and involving the founders.

The territorial menagerie

For the political geographer the southern Africa of the 1890s, a colonial realm, was a territorial menagerie (Figure 1). Two self-governing British colonies, the Cape and Natal, faced the coastline; landlocked, the Orange Free State and the South African Republic – colloquially known as the Transvaal – were republics subject to a vague British suzerainty; north of the Limpopo River a British private company with its head office in London administered the two Rhodesias; in the British protectorates of Basutoland and Bechuanaland and in the British-Transvaal condominium of Swaziland faint shadows of black kingdoms still held sway; along the eastern littoral the Portuguese colony of Mozambique stretched southwards from the Rufiji River, past the Zambezi and Limpopo deltas, past the strategic Delagoa Bay, to the Natal border; along the west coast north of the Orange River the Germans were struggling to impose their authority over South West Africa, formally a protectorate but in practice a colony.

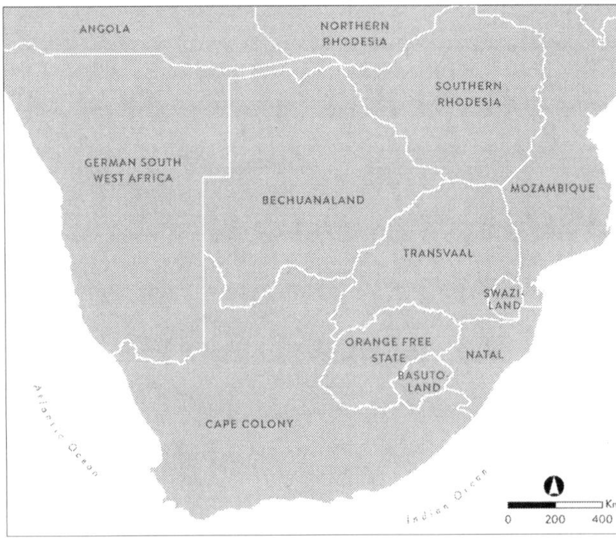

Figure 1: Political geography of southern Africa at the end of the 1800s

The discovery of gold on the Witwatersrand in 1886 rearranged the territorial mosaic around a new economic hub on the Highveld 500 km into the interior. Nine years afterwards, railways from the Cape, Natal and Delagoa Bay converged on the Rand and the South African Republic surpassed the Cape in economic output. For the British Empire builders of the 1890s the resource-rich Transvaal appeared to be the logical core of a future British Greater South Africa which promised to stretch all the way from the Cape to the Lake Tanganyika (Buchan 1903). The South African Republic, however, had its own dreams of sovereign independence. By drawing 450 000 troops from all over their Empire, Britain took the two republics out of the political game during the South African War of 1899–1902, the zenith but also the turning point of British imperialism (Omissie & Thomas 2002). The war left the subcontinent scorched and most of its people, whatever side they took, disillusioned, but the political geography was simplified: the conquered South African Republic and Orange Free State were now also British colonies and, in fact, became self-governing by 1907.

Southern Africa, neither a despotic empire like India nor an open white settlement frontier like Canada or Australia, was unique in the British imperial system. In 1904 nearly four fifths of British Southern Africa's 5,7 million people were blacks and they differed from Native Americans and Australian Aborigines by being prolific, adaptable and resilient. Practically all whites, one fifth of the population, lived in colonial states where they held power over blacks, but they were split between English speakers and Afrikaans speakers, the former urbanized, with 90% of the latter living in rural areas. Encouraged by the Liberal landslide in the British election of 1906, South African whites, conscious of being a minority and in a honest effort to bridge past differences, started exploring ways towards a more comprehensive unity. Metropolitan Britain encouraged these efforts: a self-governing South Africa was an excellent counterfoil to the threat against imperial communication lines from German African colonies (Hyam 1972).

In 1910, by consensus between the two white elites, the Cape, Natal, Orange River and Transvaal colonies merged into the Union of South Africa under a tight constitution. The South Africa Act of 1909, a measure of the British Parliament, placed the Union on par with Canada, Australia and New Zealand, self-governing under its own laws but subject to the British Crown and not yet sovereign, a formula open to different interpretations. South Africa, as opposed to southern Africa, for the first time acquired a clear-cut territorial and political identity with the prospect of strong, efficient government. Even so, for its conceivers the Union remained an incomplete territorial construct. Article 151 of the South Africa Act provided for the entry of the Rhodesias and Article 152, subject to guarantees on black land rights, for the three protectorates.

White man's land, black man's land

Political discourse during the time of Union followed ethnic and cultural divides with English-speaking whites, Afrikaans-speaking whites and blacks all conversing on different themes and with different emphases. English speakers, the real victors in the South African War, looked forward to the fulfilment of Union in a Greater South Africa whereas Afrikaners were redefining their own identity. They held one myth in common: South Africa was a white man's country and would remain one (Giliomee 2003). Some optimists even believed that white numbers would catch up with blacks by the end of the twentieth century (Muirhead 1909). Blacks lived in the fragmented remnants of the old traditional kingdoms and were deeply divided in culture and language. Politically they were powerless, their voice was muted but they knew they were South Africans. How did these three discourses develop over the next thirty years?

Until the mid-1920s English-speaking whites laid the agenda of the country's political discourse, controlled the economy, determined social norms, dominated the institutions of culture and science, and drew confidence from their links with the metropolitan core. For them Britain, where many family relations still lived, was "home". Yet, the crude imperialism of the South African War was no longer tenable. English-speaking whites accepted the obligations the imperial connection brought but regarded self-government as a right. In their eyes they were the Trekkers of the twentieth century: pioneers of trade, mining and scientific farming, looking north to Zululand, the Lowveld and beyond the Limpopo to Rhodesia, that "conquered but not entirely tamed frontier" (Dubow 2002: 86).

While English-speaking whites took a disparaging view of Afrikaners – "wretched, disloyal, unfaithful" according to a letter writer in the *Eastern Star* (Tamarkin 1996: 43) – they were open to individuals respecting the British connection and sharing their vision of a Greater South Africa. Such Afrikaners were Louis Botha (1862-1919) and Jan Christiaan Smuts (1870-1950), successive Union prime ministers between 1910 and 1924. South African troops under Botha occupied German South West Africa in 1915, the territory was ceded to the Union by the League of Nations to be administered as a C mandate and regarded as a future fifth province. Smuts' ambitions were even more spacious. Taking South African troops with him in 1916, he commanded the British campaign in German East Africa, hoping to barter the south-eastern corner of the territory with the Portuguese for Delagoa Bay. In 1923 he vainly tried to persuade Rhodesians to join the Union and during his election campaign of 1929 conjured the vision of "a great white Africa along the Eastern (highland) backbone with railway and road connections connecting North and South" (Hancock 1968: 203). But Union never expanded beyond the original four colonies.

Afrikaners took a different view of Union and Empire. Impoverished and marginalized by the South African War, the Union brought Cape, Free State and Transvaal Afrikaners within one political framework and made them aware of a common destiny. Severed from their Dutch, German and French roots for more than a century, they cherished no sentiments about a European "home". South Africa was their country, Afrikaans their language and dominion status a staging post on the way to a sovereign independent republic. As Empire changed into Commonwealth during the First World War, British metropolitan leaders became more willing to concede to the constitutional qualms of Afrikaners like Barry Hertzog (1866-1942), prime minister between 1924 and 1939, and also Canadians like Mackenzie King. The Statute of Westminster of 1931 put South Africa and its sister dominions on the same footing as the United Kingdom: "equal in status, in no way subordinate to one another", a sovereign independent monarchy under a monarch who happened to be resident in London. For many Afrikaners this measure fell short of their goal.

A second Afrikaner narrative emerged during the first decade of Union: that of poverty. As agriculture became commercialized during the early decades of the century, a fault line opened between rural rich and rural poor and, by 1930, depression and drought trapped 250 000 Afrikaners, one in four, in a class of poor whites, "ignorant, lazy, shiftless and prejudiced" (Huntington 1933: 45). English speakers with a per capita income double that of Afrikaners were hardly touched. But the United Party leadership of the 1930s faced the problem squarely and responded with rehabilitation programmes exclusively aimed at whites: "civilized" labour on the railways, irrigation schemes in rural areas, parastatals in the cities and, above all, better social and educational services. By 1940 the poor-white problem had all but disappeared and social engineering had made its mark. That blacks suffered even more, was noticed by only a view prescient whites.

Afrikaners had ambivalent views about their country as territory: they were, in fact, Little South Africans. Although Kruger's South African Republic glanced covetously at Khosi Bay, the Afrikaner Bond had doubts about annexing British Bechuanaland and Pondoland to the Cape Colony in the 1890s and, following 1910, most Afrikaners were unenthusiastic about Rhodesia. English-speaking white Rhodesians could, after all, cancel their growing majorities out at the polls. A National Party brochure during the 1929 election cited another reason for rejecting Smuts' Greater South Africa, namely that "black Kaffir state ... of which South Africa is to form such a small part that she would know her name no more" (Warhurst 1984: 94). But Afrikaner Little South Africa did go beyond Union. By 1921 South West Africa had already more Afrikaner than German residents and in 1936 Hertzog's cabinet, in the face of the Third Reich's demands for colonial restitution, committed itself to the territory's status quo (Du Pisani 1988). Both Smuts and Hertzog also tried to draw the three protectorates into the Union, Hertzog with markedly less enthusiasm than Smuts (Hyam 1973).

As the grip of the Great Depression slackened by the late 1930s, industrialization and its social consequences transformed South Africa into a country with new problems, stresses and animosities. With a population of 19,6 million in 1936, the white share stabilized at 21% but Afrikaners now had a clear majority. While white English speakers still idealized the Rhodesian frontier, they shared the sports fields with Afrikaners, accepted bilingualism and tactfully refrained from referring to Britain as "home". Like the Afrikaners they were becoming aware of their own identity: they were "Dominion South Africans" (Lambert 2000). On their side many Afrikaners discovered that the monarch was too far away to matter and Commonwealth membership had real economic benefits. This frail convergence was rudely shattered when the United Party split in 1939 and South Africa, under the leadership of Smuts, joined its Commonwealth partners in the Second World War.

As the political discourses heated up during the 1940s, two new voices rose above the din. One was that of a virulent Afrikaner nationalism to which we will return in Chapter 5. The second was the black majority who for the first time spoke for themselves. According to the 1936 census, South Africa's black population nearly doubled from a low 4,1 million in 1904 to 7,1 million. Seventeen per cent were urbanized, the permanent urban dwellers already exceeded the number of urban migrants and the discourse of black-white relations rose steadily towards the top of the agenda. In effect there were two discourses: that among whites on the political status of blacks and that among blacks themselves.

During the first two decades of the twentieth century whites thought of blacks as the "native problem", people at best too immature to fit into a modern society, at worst barbaric. The independence of the traditional black kingdoms was extinguished one after the other during the latter half of the nineteenth century and all that remained in 1910 were scattered "locations", away from the major economic nodes, and like the three protectorates, economically unsustainable. But the white commercial economy depended on black labour and blacks, in one way or another, had to be accommodated within the existing political order. Whites made the rules. The Cape constitution of 1853 placed white, coloured and black males on a common voters roll and qualified the right to vote in terms of education and property. Natal also claimed a colour-blind franchise but with a threshold so high that only one black (and no Indians) had voter's status by 1899. The Boer republics had one clear-cut principle: "no equality between White and Non-White in Church and State." They stood their ground when they became self-governing in 1907 and regarded the principle as non-negotiable in the bargaining which lead to Union.

Although some of the founders cherished the hope that liberal Cape values would eventually prevail Union-wide, northern sentiments proved to be more durable. The signposts for the future of blacks had already been staked out in 1905 by the Lagden Commission on Black Labour during the high tide of Milnerian imperialism. "Native development", the Commissioners declared, "should, firstly follow a course separate from that of whites with a 'dividing line between semi-barbarous and civilization' and should, secondly, ensure a voluntary but controlled flow of cheap labour to urban areas and the mines" (Legassick 1995: 50). Three measures of Union followed Lagden's signposts and graded them into a road. The South Africa Act of 1909 entrenched the racial status quo by limiting the colour-blind franchise to the Cape and barring people of colour from membership of Parliament. The Native Lands Act of 1913 prohibited black freehold ownership outside the reserves (only 7% of the country's area) and blocked individual blacks from becoming commercial farmers. Finally, the Native Urban Act of 1923 deemed blacks as temporary urban dwellers, required them to carry passes and denied them property rights in towns.

While the white political leadership was uneasily aware of the growing restrictions on blacks, its efforts to compensate were half-hearted. During the time of Union a buzzword from the American South entered the local political discourse: segregation, vaguely defined as a system of separate institutions for whites and blacks, the black institutions to be developed under white tutelage. Segregation was no policy, rather a slogan "drawing different groups into discourse, always promising, never quite revealing ... with as many meanings as anyone could want" (Dubow 1995: 166). Both Smuts and Hertzog paid lip service to it: Smuts emphasized the strengthening of institutions like the Native Representative Council, Hertzog wanted more black land (Dubow 1995). Some experts held the three protectorates as examples of successful segregation.

In the late 1920s, Hertzog's National Party tried to give segregation a positive image by advocating enlarged reserves as bastions for threatened black cultures, each within its own territorial framework. Initially, Hertzog won the sympathy of white liberals but their expectations, just like those of blacks, were rudely shattered by the Native Trust and Land Act of 1936. While the Act provided for the expansion of black reserves to 13% of the country's area, black voters in the Cape were removed from the common role. Coloured male voters still had reprieve for another 20 years, but segregation lost its credibility. By the early 1940s the discourse of black-white relations became harsh. "Isolation has gone", Prime Minister Smuts lamented in 1942, "and segregation has fallen on evil days" (Smit & Booysen 1981: 24).

As these events unfolded, blacks found a voice of their own. Initially, their spokesmen were educated blacks who positioned themselves away from the traditional leaders. Most of them studied at Eastern Cape mission schools and took middle-class jobs in the church, education and the lower bureaucracy. Their attitude towards white-dominated society was ambivalent. On the one hand, they assumed that their Christian education would provide access to a common South African society and, on the other hand, they hived off to separatist religious movements like the Ethiopian Church of the 1890s and the more enduring Zionist movement of ten years later. Affronted by slogans like "White Man's land" and "White South Africa", they distrusted Union and went to London to protest against the South Africa Act. "The Union of South Africa", wrote the first black historian some years later, "was welcomed by white South Africa as the beginning of a new and enlightened era. To the coloured peoples it was the death knell of a long tottering identity" (Molema 1920: 246).

The earliest black political organizations arose separately within each of the Cape, Transvaal and Orange River colonies but Union brought blacks, like Afrikaners, together in one political framework. An early consequence was the founding of the South African Native Convention in 1912, renamed the African National Congress (ANC) eleven years later. From its inception the organization tried to transcend ethnic divisions: a delegation which travelled to London in 1914 to express dismay at the Native Lands Act, consisted of two Zulus, one Xhosa, one Pedi and a Tswana. The Convention's aims were modest and phrased in the rhetoric of white hegemony: improved links with the government, black upliftment and redress for just grievances (Giliomee & Mbenga 2007). Some individuals were even willing to consider territorial partition provided the outcome was based on direct negotiations between black and white and would lead to fully autonomous successor states equally divided in terms of size and resources (Molema 1920).

During the 1920s the ranks of educated blacks were swelled by a new generation of teachers, academics, doctors and some lawyers but the ANC stagnated under indifferent leadership. Dissatisfaction with Hertzog's bills provided blacks with a cause to rally around and infused new vigour into the debate on black-white relationships. Black voters represented only 1,2% of the South Africa electorate, but the old Cape franchise had high symbolic value and always held the promise of better things to come. In December 1935, 400 delegates met in an "all African convention" in Bloemfontein to reject its abolition, the most representive assembly of black leaders up to that time. The language of their resolutions was urgent and assertive: away with white trusteeship, revise franchise qualifications, extend the Cape franchise to other provinces (Dubow 1989). The base was laid for the more militant ANC of the 1940s and within ten years black-white relations were at the top of the political agenda.

The demands of the black leaders were formulated within the legal and territorial framework of the Union of South Africa. Unlike whites they had no need to demarcate themselves against the rest of Africa, neither did they have visions of a Greater South Africa. Their struggle was for civil and political rights where they were. They were also Little South Africans.

South African science attains maturity

The preconditions for a South African discourse of geography were developed by the much more comprehensive discourse of science, always to a certain extent entwined with political discourses, timid at first and dependent on the European core but eventually attaining maturity through themes, institutions and events of its own. This section follows the discourse until the 1930s.

The beginning, of course, was indigenous knowledge, the local awareness of southern African landmarks, plants and animals, crucial for the survival of the subcontinent's communities, and often passed on to neighbours and successors. But to become science, knowledge requires systematic data collection, conceptualization, critical assessment and diffusion over an extensive culture area.

For the three centuries after 1500 the cultural core to which data on southern Africa were reported and converted into science, was Western Europe. On their way to the East during the sixteenth century, European navigators charted and named the subcontinent's coastline and their observations were reprocessed by the great map-makers of the seventeenth century (Brown 1950). During the eighteenth century scientific travellers such as Thunberg (1772-74), Sparrman (1775-76) and Masson (1772-74; 1786-95)[1] ventured into the Cape's interior to collect plants and animals, referring them to European universities and botanical gardens to be taxonomically classified (Forbes 1963). In due course reports on observations in South Africa were called upon to settle scientific controversies in the European core. In 1752 Abbe de la Caille's measurement of the arc of the meridian between Cape Town and the present village of Aurora upset an earlier conclusion by the French Academy that the Earth was flattened at the poles (Raven Hart

1976). De la Caille's measurement was later found to be erroneous (Gill 1905). Early in the nineteenth century the British geologists Playfair and Lyell used descriptions of granite veins in the slates underlying the sandstones of Table Mountain to prove that granite was an igneous rock. Their interpretations were confirmed in 1836 by Darwin's brief but accurate interpretation of the Sea Point granite contacts (Barnard 2004).

Although the European core remained the centre of reporting, a local scientific discourse became apparent by the mid-nineteenth century. Following the support of older institutions such as the Royal Observatory (1820) and South African Museum (1825), the Cape colonial government started employing professionals, at first precarious in their tenure and then on a longer-term basis: a Conservator of Forests in 1847, a government geologist in 1855, a botanist in 1858 and a Meteorological Committee in 1860 (Beinart 2003). Moreover, the expansion of schools and hospitals brought educated teachers and doctors into the country who were fascinated by the local environment and speculated about it in journals like the *Cape Monthly Magazine* (1857-1876).

The emerging discourse drifted in two opposite directions. Medical men lauded the healthy Karoo, and botanists and zoologists, the pure scientists, were enthralled by the country's richness in species. Applied scientists, on the other hand, were overawed by problems posed by vermin, stock disease and soil degradation. Aridity was an even more enduring threat. Already in the 1820s Karoo sheep farmers complained that rainfall was diminishing and springs drying up. When Anderson (1850-66), Chapman (1854-64) and eventually Livingstone (1853-71) expanded the frontier of known southern Africa beyond the Cunene and Zambezi Rivers, the dry lakes and fossilized stream channels of the Kalahari seemed to suggest that desiccation was a process of subcontinental extent still in progress. In a lecture before the Royal Geographical Society, Wilson described these telltale features and put the blame on overgrazing, veld burning and deforestation. For Livingstone, on the other hand, the culprit was recent crustal movements. The popular and scientific discourses converged: "South Africa was drying up!"

Geologists working in southern Africa also discovered that the subcontinent's stratigraphy presented difficulties of its own. There were, to begin with, rock types unknown to Europe. The glacial origin of Dwyka tillite was contested from the 1850s to the 1890s and the formation was burdened with bizarre names like "claystone porphyre, trapped conglomerate, metamorphic conglomerate, melaphyre breccia and so forth" (Eales 2007: 118). The stratigraphic succession posed similar problems. The pioneer geologists of the 1840s and 1850s vainly tried to fit southern African formations into the standard European column, until RJ Dunne, in his map of 1877, broke with northern hemisphere models and introduced the local names still in use: Karoo, Cape, Nama and, after many misplacements, Transvaal (Anhaeusser 1997). In 1904 Corstophine collated the diverse regional geologies into one standard stratigraphical column and in the following year, in collaboration with Hatch, published the first South African geology textbook (Hatch & Corstophine 1905). South African geologists also agreed on two other points: local formations were much older than those in Europe and, especially because of the common presence of tillite, displayed a remarkable similarity with those of India, Australia and South America.

The turn of the nineteenth century brought new institutional bases for scientific discourse. The technical services of the colonial governments continued expanding and their staff became more specialized. Once again the Cape took the lead but by the time of self-government in 1907 Transvaal was not far behind. As the professionals sought contact with their peers, the first learned societies were established: some focusing on a basic science (e.g. the Geological Society of South Africa, 1895), others taking care of the professional interests of their members (e.g. the Transvaal Medical Society, 1897) and a few trying to bring together specialists from different fields. The discourses engendered by these societies were formalized in journals and events such as conferences, public lectures and excursions.

The earliest learned society was the Cape Town-based South African Philosophical Society, established in 1877, comprehensive and elitist to the extent that the exorbitant entry fee of £2 per annum was levied to restrict membership to 80 persons! In 1906, by royal consent, the Society renamed itself the Royal Society of South Africa and its *Transactions*, commencing in 1878, is still being published, the oldest scientific journal in the country (Brown 1977). As the development of the Witwatersrand gold mines brought the infrastructure of industrial capitalism to South Africa, most of the professional staff took up residence in Johannesburg and the majority of learned societies found it convenient to have their administrative offices there. Their membership was nearly exclusively English speakers. The British Empire stood for science and science was progress (Dubow 2006).

As the South African War was winding down, Sir David Gill (1844-1914), Astronomer Royal in Cape Town and South Africa's senior statesman of science, took the initiative to establish a broad-based society, modelled on the British Association for the Advancement of Science, which would appeal to both professional scientists and the informed public. His efforts crystallized into the South African Association for the Advancement of Science (S_2A_3). The inaugural conference was held in Cape Town in April 1903 and the society initially embraced all fields of learning, from the natural sciences to linguistics and philosophy. Within two years the Association's membership exceeded 1000 and members hailed not only from the four colonies but also from Rhodesia, the protectorates, Mozambique and German South West Africa. The first annual conference met in Cape Town and then in Johannesburg, eventually wandered out to Kimberley, Caledon, Outdshoorn, Lourenço Marques (Maputo), Bulawayo and others. Subcontinental in extent, the Association seemed to be the scientific precursor of Union, if not of Greater South Africa.

Institutions prompted events. One of Gill's ambitions was to link the S_2A_3 with the British Association by hosting a joint conference in South Africa. The event came to pass in August and September 1905 with the generous support of the colonial governments, the first time the British Association met outside the United Kingdom. Among the 200 foreign participants voyaging to South Africa with discount from the Union Castle Line were some of the most famous scientists of the day – the Darwins, Rutherfords and Gregorys – and they were responsible for the bulk of the papers, but a few South Africans made modest contributions in geodesy, astronomy and geology (Anonymous 1906). The sessions in Cape Town and Johannesburg were followed by excursions, receptions and postprandial speeches. The dignitaries' message was loud and clear: universal science would bring healing and progress to a devastated country but South African science should be linked with the imperial core. Fittingly, the climax of the conference was the opening of the Zambezi Railway Bridge by Prof George Darwin, president of the British Association, Cambridge astronomer and mathematician, and son of the illustrious Charles Darwin (Dubow 2006).

Although still peripheral to the European core, South African science was gradually acquiring an individuality. Union in 1910 was an important catalyst. Whatever its shortcomings, the new state was more efficient and wider in reach than its colonial predecessors and it had a keen appreciation of the subcontinent's environment. New research addressed both national problems such as drought and irrigation and regional problems such as bovine parabotulism (*lamsiekte*) in the Kalahari and ostrich breeding in the Karoo. Government services were expanded to collect and disseminate information on subcontinental scale: a Geological Survey in 1910, a Meteorological Office in 1912 and a Bureau of Census and Statistics in 1917, the head offices all centralized in Pretoria.

By 1920 the reports of these offices provided data useful for assessing and managing environmental problems. In 1912-13 and again in 1918-19 South Africa was stricken by ruinous droughts and in 1920 the Union Government appointed a high-powered commission to enquire whether South Africa's rainfall was indeed diminishing, as popular myth would have it. From the relatively short runs of rainfall data obtained from the Meteorological Office, the Commission (U.G. 49/23: 6) found "no proof of an appreciable change in the average annual rainfall during recent historical times", but harmful farming practices undermined its economic value. Rainfall variability and climate change were to have a long refrain in the geographical discourse.

A controversial issue in the early years of Union was the reform of tertiary education which since 1873 was managed, to everyone's dissatisfaction, by the University of the Cape of Good Hope. By 1910 the university laid down curricula, took examinations and awarded degrees but left teaching in the hands of some ten colleges, from the South African College (SAC) in Cape Town to the Transvaal University College (TUC) in Pretoria. Union inspired some English South Africans with the vision to transform the old University of the Cape of Good Hope and its colleges into one national teaching university on the Oxford model and were promised generous financial support by some Randlords. The vision, however, became clouded. The 1891 will of Cecil John Rhodes, the diamond tycoon, provided land and funds for a university at Groote Schuur along the slopes of Devil's Peak, while the bequests of Beit and Wernher aimed at a university on the Witwatersrand, eventually pitting the South African College against the South African School of Mines and Technology. In its in-between position, the Victoria College at Stellenbosch – "that happy nursery of Hertzogism and reaction" according to the Johannesburg *Star* (Boucher 1974: 148) – felt its Dutch-Afrikaner heritage threatened.

The Union Government compromised by decentralization. The three university acts of 1916 awarded full university status to the South African and Victoria Colleges, in 1918 becoming the Universities of Cape Town (UCT) and Stellenbosch respectively, while the other colleges united in a federal University of South Africa with its administrative head office in Pretoria and with clear guidelines for eventually attaining university status. The acts allowed all present and future universities to conduct their own examinations, confer differentiated BA, BSc and BCom degrees and organize their teaching within departments responsible for a particular field of knowledge. Among the fields considered was geography.

Within ten years South African universities diverged according to the local community and its language of instruction, reflecting the duality of white culture. Cape Town and what became Witwatersrand, Rhodes and Natal Universities never seriously considered any other language than English and, through the interchange of staff and students, forged creative links with British universities. Their practice of science and arts, they claimed, was universal and "open". Stellenbosch, and later Pretoria, Bloemfontein and Potchefstroom, opted for Afrikaans and accepted responsibility to develop it into a language of science. Their links with continental and British universities, although not negligible, were tenuous and they were often accused of Afrikaner particularism. They did not admit students of colour.

The extended government services and the reformed tertiary institutions profoundly affected the discourse of science. Between 1905 and 1925 the number of university teachers increased tenfold and the annual number of graduates twelvefold while the number of scientific societies nearly tripled (Hofmeyr 1929). Universities and societies became local centres of dissemination, the papers published were no longer reports on South African marvels for European readers but a sincere effort to contribute to the universal body of science. And in the years before "big science" South Africans were highly successful. The UCT anthropologists Radcliffe-Brown and Schapera applied structural functionalism to the interpretation of traditional societies and equipped the discipline with an analytical tool which remained in use until the 1950s (Hammond-Tooke 1997). The discovery of the Taung skull in 1925 left Raymond Dart and Robert Broome with the suspicion that *Homo sapiens* originated in Africa, while Alex du Toit, "the world's finest field geologist" (Mesnard 1986: 82), in an effort to explain the similarity of rocks in the southern continents, placed southern Africa in the centre of a sundered and drifting Gondwanaland. Time proved both Dart and Du Toit correct but initially the European scientists had nothing but contempt for their ideas. Du Toit's (1924: 59) rejoinder was scathing: "The European aspect has become unduly emphasised, with a tendency...to a somewhat narrow outlook on what are really world-wide problems." Thirteen years later he went further: "Their criticism of drifting continents...is only partly justified, trivial and even fatuous" (Du Toit 1937: 6).

The new confidence was displayed in 1929 when the S_2A_3 and the British Association held a second joint conference in South Africa, with Cape Town, Johannesburg and Pretoria as venues. Participation was even greater than in 1905: 500 scientists from all over the world, ten official guests from the United Kingdom, Netherlands and France – the countries of white origin – and some 75 southern Africans presenting papers, mostly on anthropology, botany and agriculture.[2] The full panoply of excursions and receptions was repeated but the rhetoric was different: South Africa itself was now the focal point. The presidential address of the S_2A_3 read in Cape Town on 22 July 1929 by Jan Hendrik Hofmeyer (1894-1948), a previous principal of Wits and just retired as Administrator of the Transvaal, affirmed this confidence:

> (We) brought to science the distinctive features of the South African outlook – freshness and breadth of view, receptivity to new illuminations and readiness to use old truths in the light of wider bearings....[And in] the mind of the nation there is being developed a new conception of a South Africa which consciously and deliberately seeks to play its part on the African continent, not aiming at conquest or domination but giving its intellectual resources to all who are engaged in developing this area of the Earth's surface" (Hofmeyr 1929: 8).

Once again the discourses of science and politics entwined but with the political Greater South Africa recast into a Greater South Africa of science. Ironically, Hofmeyr looked northwards but there was only one black among the 75 South African speakers, viz. Dr DDT Jabavu (1885-1959) of Fort Hare. Fifteen speakers were Afrikaners and there were also three geographers: Serton of Stellenbosch, Wellington of Witwatersrand and Plummer of Pretoria.

Geography as counterpoint

Participation in 1929 signalled the entry of South African geographers into the local discourse of science and also into the discipline's international network. Even so, geography had a long prehistory which formed a minor but distinct counterpoint to the emergence of South African science.

While Willem Janszoon Blaeu and Jan Jansson were including maps of southern Africa in their atlases, numerous seventeenth century travellers to the East put their experiences in writing and devoted some cursory pages in their voluminous books to their brief stays at the Cape. The texts were travelogues meant to be read for pleasure: unsystematic, intensely personal, directed at the exotic, everything seen through European eyes and filled with anecdotes on ascents of Table Mountain and encounters with the Khoi (Raven Hart 1971). The first half of the eighteen century brought observers who stayed longer and learned more. Peter Kolbe's *Complete Description of the Cape of Good Hope* (1719) lost credibility over time but remains a valuable source on the Khoi. In turn, Otto Friedrich Mentzel's *A Geographical and Topographical Description of the Cape of Good Hope* (1785, 1789), based on his residence there between 1733 and 1741, provided lively descriptions of the burghers. Mentzel's calling his text a geography is an exception. Most visitors eschewed the term but provided information for the geographers of the European core.

As the descriptions of sojourners at the Cape made way for scientific travellers during the second half of the eighteen century, their reports became concise, systematic and minimalistic. Georg Forster (1754-1794) and his father Johann Reinhold were naturalists on the Resolution during Captain Cook's circumnavigation. George devoted Chapter 3 of his *A Voyage Round the World* (1776) to the Cape where the ship had lain at anchor for three weeks in February 1875. After the usual anecdotal introduction, Forster reviewed in logical sequence the administration of the Cape settlement, the burghers, the slaves, agricultural activities with special emphasis on viniculture and, finally, a brief summary of animal life. His vision was impersonal and factual, his language that of the Enlightenment (Pakendorf 2003). "Brothers", he proclaimed later in life, "you are here to search for the truth" (Steiner 1985: 20).

Forty years later Forster's approach was taken a step further by Carl Ritter (1779-1859) who, between 1820 and 1859, was professor of geography at the prestigious University of Berlin. Ritter was a geographer by choice and saw his task as demonstrating the unity among the diversity of Earth-surface features within a particular area. Humans, he argued, should be placed against the background of the natural environment where they lived, the broad pattern of the particular area should be identified and the relations between these features and human history should be followed up. Ritter applied his outline in three chapters on southern Africa in Part 1 of his *Earth Science* (1822). He summarized the relief of the subcontinent in the units which all subsequent textbook writers have followed: a central plateau bounded on the outside by a steep escarpment, marginal lands rapidly descending westwards and eastwards toward the Atlantic and Indian Oceans respectively and a more gradual descent southwards across the terraced lands of the Great and Little Karoo to the southern littoral. From this base he discussed the region's ethnically-diverse communities and their stock farming. Ritter's presentation had its defects. In contrast to Forster he was an armchair geographer who never saw southern Africa: Ritter's sources were Barrow and, to a lesser extent, Thunberg, Liechtenstein and Latrobe and the relationships he emphasized fell short of explanation.

For South Africans, texts aimed at the European core were models to be admired from afar. A geographical discourse however depended not only on books but also on institutions which were slow to emerge.[3] Geography started as a lowly primary school subject in the Cape curriculum of 1839, supported in due course by textbooks linked to maps and overflowing with place names to be mastered by rote learning. Although the University of the Cape of Good Hope allowed a niche for the subject at secondary level, few schools availed themselves of the opportunity and the subject drifted aimlessly into the twentieth century. "The official opinion in South Africa", mourned the Johannesburg *Star*, "is that Geography is a study neither suitable nor necessary for boys above the age of 12" (Quoted by Flowers 1906: 213).

The 1905 conference of the S_2A_3 and British Association marked a new beginning. The organizers referred papers they regarded as geography to Section E, a catch-all forum where army officers could reminisce about their expeditions and surveyors could display their maps. The opening paper was read by Rear Admiral Sir WJL Wharton KCB FRS, although among the participants were some of the most famous names in contemporary geography: the geomorphologists William Morris Davis of Harvard and Albrecht Penck of Vienna and the two Englishmen, Hugh Robert Mill (1861-1950) of London and AJ Herbertson (1865-1915) of Oxford (Anonymous 1905).

In their conference presentations and, more pointedly, in their papers following the excursions, Davis (1906a, 1906b)[4] and Penck (1908a, 1908b) considered and analysed Passarge's just published explanations of southern African pans and inselbergs, sometimes in disagreement. Geomorphological themes also surfaced in Section C (Geology) where Rogers, Du Toit and Schwarz presented papers. The conference and its repercussion introduced South African landforms to the Earth sciences and involved Passarge and Schwarz, the first two founders treated in this book, in themes which overlapped with geography. During 1905 Passarge was called to a chair of geography at Breslau University in Germany (today Wrocław, Poland), three years later his book, *Südafrika*, was published in Leipzig and found a respectful readership in South Africa.

Institutionally, the subject was also built from school level upwards. At the 1905 conference AJ Herbertson presented a paper on his new map of African rainfall but his real interest was the status of geography at schools in the United Kingdom. In meetings with Cape Town teachers, away from the formal glamour and rhetoric of the conference, he enthusiastically extolled a "new geography" based on causal explanation. His argument was taken up by young teachers in the Cape Colony, promoted in the official *Education Gazette* and *Educational News* and given urgency with the advent of Union. An early and tenacious exponent was Paul Albert Ritter (1872-1952), initially of George Boys' High School,[5] was joined in 1914 by James Marnoch Hutcheon of SAC in Cape Town. Geography at school, they claimed, seemed "dull and senseless" only because teachers were not trained in the "new geography". What the subject should be, Hutcheon (1916: 329) contended, was "the study of man's 'passive' and 'active' relation to his natural environment." Geography's educational excellence lay in thrashing out these relationships.

In the early 1900s the clamour for an improved school geography was heard in the higher corridors of education but where were the teachers to come from? Two university colleges responded. In 1913, prompted by a small group of its Scottish professors, Victoria College at Stellenbosch decided to offer a one-year course in geography and obtained the University of the Cape of Good Hope's recognition of the subject for the BA Pass examination (Wesso 1992). The course was implemented in 1914 and teaching fell to the professors of geology, zoology and history, not always acting in perfect equanimity. In 1917 the South African School of Mines and Technology (after 1922 Witwatersrand University) went one step further by appointing James Hutcheon from SAC as full-time lecturer.

Geography's final acceptance came in 1918 when the Joint Matriculation Board recognized the subject for matriculation, and university colleges, to meet the demand for teachers, rushed to include geography in their curricula. School and academic geography converged. The larger institutions opted for full-time specialist geographers who of necessity had their training overseas, and soon offered the complete hierarchy of courses, from first-year to master's level. The smaller institutions combined geography as junior partner with a cognate discipline, usually geology, and took more time to achieve maturity. Among the specialists a young Dutch academic Petrus Serton was appointed at Stellenbosch in 1920, the Englishman John Harold Wellington at Witwatersrand in 1922 and Francis Plummer, another Englishman, at TUC (after 1930 Pretoria University) in 1926. UCT dragged feet until 1936 when William Talbot took up a chair. The smaller Rhodes University College established a one-year course in geography in 1919, Grey University College and Pietermaritzburg University College followed in 1919 and Potchefstroom in 1927. Just as at Bloemfontein and later Potchefstroom the subject geography at Rhodes was assigned to the professor of geology.

Just as in countries of the European core, the university geography departments transformed South African geography into a professional discipline. Developing a course structure impelled each founder to structure, articulate and apply his vision of the discipline, while research students moved out and formed a supporting network in which new insights were developed and fed back to the founder to be weaved into courses and publications. Publications, in turn, conferred status on the author as well as his institution. With the exception of Stellenbosch, the contents of the early courses stayed close to Herbertson's deterministic model of fifteen years before and reflected Anglo-Saxon thinking of the time. Geography was supposed to study the influence of the inorganic environment on life and especially humans, physical and regional geography were the leading sub-disciplines and most graduates went into teaching (Wesso 1992). Soon papers covering themes on southern African geomorphology and climatology appeared in journals as widely separated as the *South African Journal of Science* and

Petermanns Geographischen Mitteilungen but, apart from some German studies on South West Africa, they did not present themselves as geography. They nonetheless supplied source material for "that little gem", Alex du Toit's *Physical Geography for South African Schools* (1912) and Ernest Schwarz's *A South African Geography* (1920), both school handbooks not tied to a curriculum.

Institutionalization at universities called a community of geographers, perhaps 30 in number, into being and the annual Pretoria meetings of the geography representatives from the constituent colleges of the University of South Africa provided a formal focus for discourse. But more was needed. During his first year at the South African School of Mines and Technology the enthusiastic Hutcheon opened a parallel route by establishing the South African Geographical Society (SAGS). Taking the Royal Scottish Geographical Society as model, he wanted to rally researchers, scholars from related disciplines, teachers and the informed public around the themes of maps and discovery. His vision was "popular" rather than "professional". Not surprisingly there was only one geographer among the Society's first four presidents: a geologist, a botanist and a zoologist made up the rest.

Like the university departments, the SAGS took the educational value of geography seriously and envisaged public lectures, excursions, exhibitions, the setting up of a library and the publication of a journal as future tasks. Events and infrastructure were of necessity tied to Johannesburg where the Society had its administrative office and until the 1950s sixty per cent of its members were resident on the Witwatersrand (Muller 1954), which left sceptics with the suspicion that it was too Wits-centred. But location in the subcontinent's economic hub also had advantages. Most scientific societies had their administrative head offices in Johannesburg and four university institutions were situated within a radius of 60 km. The *South African Geographical Journal* (SAGJ) was an enduring success. Published on an annual basis since 1917, the first four volumes were slim, few of the initial contributors were geographers and the contents accommodated travelogues and hobby horses together with real substance. The *Journal* survived an early crisis when, after Hutcheon's untimely death in 1921, the Wits professor of zoology had to guide volume four (1920/21) through the press.[6] In 1923 Wellington, Hutcheon's successor, took over the editorship, the *Journal*'s cover acquired a spartan simplicity, contributors were drawn from all over the Union and the contents steadily graded into professional geography (Jackson 1978).

As preparations were made for the 1929 conference of the S2A3 and the British Association, Wellington was requested to coordinate the papers and sessions in Section B, the niche allocated to geography. Again the section had to host a melange of peripheral topics, but now the core was unmistakably geography. The participating British geographers – among them Marion Newbigin (1869-1934), for many years editor of the *Scottish Geographical Magazine*, JL Myres (1869-1954), a classical historical geographer from Oxford and HJ Fleure (1877-1969), an anthropogeographer from

Aberystwyth – commanded respect and fraternized with the South Africans. The latter, in turn, targeted situations unique to their country and challenged conventional wisdom in the light of their local experience.

In Cape Town, Serton (1929) argued that the notion of desert depended on human perception. The sheep farmers of the western Great Karoo did not regard their arid environment as desert and used its resources more efficiently than ranchers in similar areas of the United States and Australia, who shunned them as wasteland. Wellington assessed physical-geographical influences on human activities in Southern Africa and knocked down the snap judgement of Ellen Churchill Semple (1911), who dismissed the subcontinent's "peculiarly despotic" climate as too monotonous for optimum human performance:

> One wonders whether the writer has ever felt the call of the wide spaces, the charm of the simple life in a genial climate, the love of freedom and independence, and the fascination of grappling with savage nature (Wellington 1929: 94).

Wellington (1932) and Serton (1936) shared the South African thrust of the 1929 conference and were soon drawn into international publications. The discourses of geography and science harmonized, the discipline attained a voice of its own and it clearly had a future.

Coming from Europe, the founding geographers joined the conversation at different times and places: Passarge and Schwarz between 1895 and 1905, Serton and Wellington in the 1920s and Talbot in the 1930s. In Chapters 3 to 5 the attention shifts to Passarge, Schwarz and Serton with a focus on each of them as individuals: personality, home background, places of study, teaching and research and, through their writing, the inevitable encounter with Adamastor. Volume 2 in the series will document the life-histories of Wellington, Talbot and King.

Endnotes

1 The dates in parenthesis indicate the years between which the travellers, from Thunberg to Livingstone, journeyed in Southern Africa.

2 The number 75 is an informed guess. Volume 26 (1929) of the *South African Journal of Science* published 79 papers of which 69 were conference presentations by South Africans. However, some, like that of Serton, were published elsewhere, in his case the *South African Geographical Journal*, No 12, pp. 26-31.

3 The paragraphs that follow on the establishment of geography as a South African school and university subject draw heavily on Wesso's detailed exposition and insights in Chapter 7 (pp. 219-251) of his doctoral dissertation.

4 During his long career William Morris Davis, the leading Anglophone geomorphologist of the early twentieth century, hardly paid attention to any other subdiscipline of geography, although a passing remark in his paper on the Cape Folded Belt in Volume 17 of the *Bulletin of the American Geographical Society* showed that he could look beyond the lie of the land. "Even to us, as passing strangers", Davis (1906b: 621) wrote presciently, "it was evident enough that the long-lasting problem of the native races was the greatest which Europeans in South Africa have to face – far greater than the

misunderstandings which caused ... the Boer War, greater than the difficulties due to lack of rainfall, and to the absence of lofty mountains and of the perennial streams that they would supply; greater than the trouble over rinderpest and other diseases of livestock."

5 Paul Albert Ritter's enthusiasm for geography was not by chance. Born in a family of German missionaries, he was the grandson of Albert Ritter of Frankfurt-am-Main, the younger brother of Carl Ritter (1779-1859), the famous professor of geography at the University of Berlin between 1820 and 1859. Paul Albert was very much aware of his family history, a summer school semester at Oxford University's School of Geography in 1907, when AJ Herbertson was at his most persuasive, probably strengthened his fervour for geography. After spending eight years (1902 to 1910) at George Boys' High School, he moved to Robertson in 1910 and Oudtshoorn in 1924 where he retired in 1931. At Robertson and Oudtshoorn he was involved in training primary-school teachers and applied the principles which he had preached in his younger years.

6 An appreciative obituary of JM Hutcheon – at the time of his death in 1921 a Fellow of the Royal Scottish Geographical Society – appeared in volume four (1920-21) of the *South African Geographical Journal*, initialled by one "WRM" Circumstantial evidence suggests that the author was William Reith McGregor, the first principal of the Johannesburg Training College for Teachers, Hutcheon's contemporary at Aberdeen University in Scotland and a member of the first board of the South African Geographical Society.

CHAPTER

3

Passarge, the First Professional

In contrast to the other founders, Siegfried Passarge spent only 31 months – from June 1896 to January 1899 – in southern Africa, never saw the subcontinent again and hardly wrote anything on its problems during the last forty years of his life. Why regard him as a founder? Passarge was the first professional geographer to visit the subcontinent and his 45 publications on southern Africa up to 1915 – among an astounding total of 113 – took a broad view of its geography: base maps, stratigraphy, geomorphology, ethnography and economic development. His works provided a systematic overview of the subcontinent's features, introduced explanations of some of them to an international audience and guided the local geographical discourse toward professionalism. That his attitude and ideas were sometimes controversial only widened the discourse.

Times and places in Passarge's trajectory were crucial to his practice of geography. The narrative commences with his formative years and education in Bismarck's Germany, concentrates especially on his southern African sojourn and the subsequent monographs, and ends with his stormy years at Hamburg University between the two World Wars.[1]

Königsberg, Jena and Cameroon: Learning the geographer's craft

Siegfried Passarge was born on 29 November 1866 in the East Prussian town of Königsberg as the son of Louis Passarge, a judge of French Huguenot descent. Königsberg, today the Russian exclave Kaliningrad, was founded in 1242 on the Baltic coast during the eastward wave of German colonization in the thirteenth century and became a prosperous Hansa port during the Late Middle Ages. The town struck deep cultural roots: from 1701 the Prussian kings crowned themselves in its cathedral, and the Protestant university, established in 1544, attracted famous scholars, among them the geographer Varenius, the mathematician Euler and the great philosopher of the Enlightenment, Immanuel Kant. East Prussia was also an area of cultural overlap where a German Junker aristocracy lorded over Polish and Baltic peasants (Rovan 1995). The Junkers were conservative, disciplined and conscious of their military tradition. The Passarge family, upper-middle class in background, shared these sentiments. Louis Passarge immersed himself in the East Prussian landscape, had wide European travel experience, loved a good story and couldn't resist lecturing his son on the difference between sturdy Nordics and capricious Mediterraneans.

At school the young Siegfried was initially an indifferent learner who shunned books for the forests and lakesides of his *Heimat*. The events of the 1870s, glowingly extolled by his father, filled him with ardour. Through his stamp and coin collections he became aware of the new German Empire, he refought the Franco-Prussian War with toy soldiers and plotted the Russian Balkan campaign of 1878 on a map. In his father's spacious library he was entranced by the adventures of German African explorers and names like Barth, Büttner, Müller and Claus echoed in his mind. "One day", he firmly resolved, "I will be an African traveller myself" (Passarge 1951: 55).

In 1886 Passarge completed gymnasium and, eager to go exploring, arranged an interview with Ferdinand von Richthofen (1833-1905), the doyen of German geographers and an experienced traveller himself (Dickinson 1989). The turn of the nineteenth century was the golden age of German geography. Because the Imperial General Staff regarded the discipline as essential background for army officers, the number of geography departments at German universities exploded from only one in 1871 to 23 by 1910, far more than any other country (James & Martin 1981). Moreover, the Empire had just come into existence and its future citizens, the children at school, had to be instilled with a clear image of their country with its diverse regions and contested boundaries, a task for the patriotic geography teacher. University senates, however, suspected that geography was being imposed from above and the first professors, mostly geologists, were frigidly received. "A geographer", ran a pun circulating in academic circles, "is someone who studies the Earth's surface, superficially" (James & Martin 1981: 72).

But the founders persevered, redefined their ancient field of study and, with the revered Alexander von Humboldt as model, laid down new norms of practice: leave your stuffy map room and do fieldwork, specialize in one systematic sub-discipline and in one particular region away from your home country, where your professional insight may be critically assessed (Beck 1973). In his famous Leipzig inaugural lecture Von Richthofen also added one other imperative: as natural scientists geographers should explain causally, a conception of geography that was accepted worldwide. In 1884 Bismarck proclaimed a protectorate over South West Africa and within ten years Germany obtained a scattered and shaky colonial empire. In the process German geographers, like their British and French counterparts, now had virgin territories available for their regional studies. No wonder that most of them became ardent nationalists. "Geography", asserted Alfred Kirchhoff of Halle, "is that most German of all sciences" (Sandner & Rössler 1994: 117).

While Von Richthofen was eventually to become Passarge's sponsor in the world of geography, their relationship at first seems to have been strained. Von Richthofen, apparently, was unenthusiastic about Africa as a research field for a young geographer (Stahlberg 1926) and his advice to Passarge was practical: begin your studies with hard natural sciences like physics and geology. After only one semester in Berlin, Passarge opted for the more sympathetic atmosphere of Jena University, just across the border in Saxony, where three Africa specialists served on the staff, two geologists and a geographer. Young Siegfried remained restless: by the middle of 1887 he moved again, now to study medicine at Freiburg on the margin of the Schwarzwald. A medical qualification, he thought, might provide entry to an expedition of discovery. At the end of 1888 he returned to Jena, crammed in geology alongside his medical studies, attained a doctorate on fossil laterites in the German *Mittelgebirge* in 1890 and passed the *Staatsexamen* in medicine two years later.

Passarge's hunch was on target. In 1893 he was invited to participate as medical officer in a semi-official expedition assigned to negotiate a northern boundary for the fledgling German colony of Cameroon. There he learned the realities of research in a remote backland: long route marches, persistent logistical problems, tropical diseases, evanescent locations, aneroid profiles, astronomical positioning and the ultimate reward: a half-empty map. After 13 months the expedition concluded its work and at home Passarge summarised his experiences in *Adamaua* (1893), an area study of 573 pages. Brimming with the self-confidence of youth Passarge described the inselbergs and duricrusts of the northern Sudan region of the colony but also displayed his prejudices. Negroes were lazy and passive, he asserted, and it would serve no purpose to Christianize them, a statement which drew immediate fire from the German missionary lobby. As a geography text, nevertheless, the book was well received in both German and British journals: "a pearl in our colonial literature" according to Von Richthofen (Passarge 1951: 141). Passarge has gained a foothold among geographers

Facing the Kalahari Thirstland

The opportunity that mattered arose in the same year *Adamaua* was published. Back in Germany from the Cameroon, Passarge settled in Berlin, re-established contact with Von Richthofen and shopped around for a job. In August 1895, while presenting a paper on Cameroon duricrusts at the London conference of the International Geographical Union, he was offered the job of assistant geologist in the newly established British West

Charterland Company and invited to join their prospecting team in a search for minerals in the far north-western corner of Bechuanaland.

The 1890s was a turbulent decade for South Africans. The Witwatersrand mines reached full production and gold shares on the London Stock Exchange bloomed but authority was vested in Kruger's South African Republic. The British South Africa Company of Cecil John Rhodes, to which the governments of the Rhodesias were chartered, went all out to discover a new Witwatersrand north of the Limpopo only to realize that Rhodesian ores were too scattered to support a large industry. Determined to make the Transvaal more accessible, the impatient Rhodes burnt his fingers in promoting Jameson's muddled raid for Johannesburg in January 1896 and had to face a Matabele revolt in Rhodesia five months later. With the shares of the British South Africa Company tumbling, some enterprising investors looked westwards to Bechuanaland, hoping that Witwatersrand sediments, famous for the consistency of their ores, might extend beneath the Kalahari sand. West Charterland was based on such expectations and took care to distance itself from the hated British South Africa Company (Maylam 1980).

Passarge disembarked from the Norham Castle in Table Bay on 3 June 1896, entrained northwards, marvelled at the successive fynbos, Karoo scrub and Highveld grasslands along the way and based himself for three weeks in Johannesburg to become acquainted with the mining industry. He viewed transport as the key to the development of the subcontinent which he partitioned into a railway South Africa, a wagon South Africa and a trackless sandveld (Passarge 1951). Only ten years had passed since gold had been discovered and the Witwatersrand towns, at the hub of railway South Africa, were physically, commercially and politically still raw, the playing field for unbridled capitalism. Passarge had no doubt who the villains were:

> I rode the thrust of a rapidly developing world in which the most concentrated work at the scientific, technical and economic level ran parallel with the wildest stock swindles and shaky speculation profits. Above all, I learnt to know the significance of the most important, influential and successful representatives of its plutocratic master layer: that of German and Asiatic Jews (Passarge 1951: 171).[2]

West Charterland's gaze was fixed on the Kalahari (Figure 2) and for many South Africans the northern Kalahari, the Ngamiland popularized by Livingstone and Andersson's travelogues, was shrouded in mystery: a thirstland with feverish swamps and teeming herds of game, a sandveld where lost cities were smothered by dunes and a succession of ephemeral lakes whose dried-up shores were littered with ivory. Passarge experienced the other side of the coin. In October 1896 the West Charterland expedition left Palapye, then still beyond the terminus of the Bechuanaland railway under construction from Mafeking, to establish a base camp at the Kwebe Hills south of Lake Ngami and 475 km to the north-west. Because of the rinderpest epidemic in 1896, the expedition had to exchange oxen for donkeys as draught animals, the route included waterless stretches of

70 to 110 km, the party plodded through the deep sand at two kilometres an hour and Passarge became infected with dysentery which trapped him for four weeks on his wagon by day and beneath the wagon at night. But he recovered and when the head geologist quit, Passarge was promoted in his place and, after arriving at the Kwebe base camp, started planning the search for minerals which was to take him all over Ngamiland. Then thirty years old, of below-average height but strongly built and completely bald, Passarge won the confidence of West Charterland's South African manager, Major Frederic (later Lord) Lugard, who described him as "the very embodiment of energy" (Wilmsen 1997: 30). Other companions found him difficult to get along with.

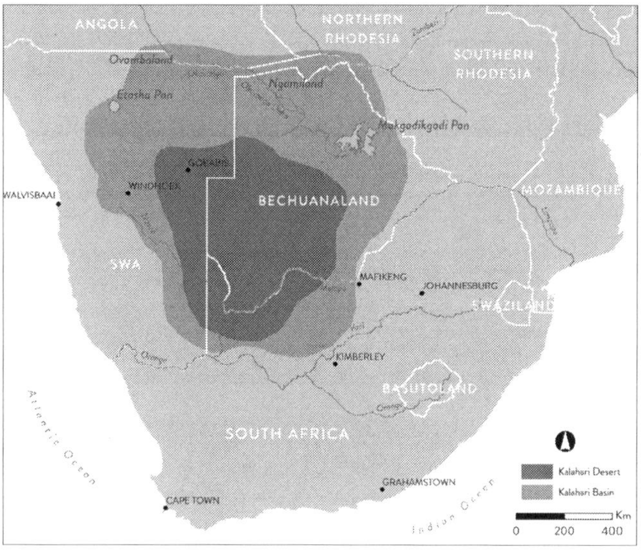

Figure 2: The Kalahari region: study fields of Passarge

While Passarge dutifully investigated numerous inselbergs for gold and traced pan depressions for diamonds he remained in his own mind a scientist rather than a prospector. Wherever he went he scrutinized the surrounding landscape and made sketches; sometimes he joined his Tswana and San drivers at their campfires and listened to their yarns; he took notes, determined the coordinates of 63 sites astronomically and compiled a base map on a scale of 1:1 000 000 which showed for the first time the anastomosis of the lower course of the Okavango River as a "delta" (Stigand 1923; Van der Post 2005). He also noted the Kalahari's rare moments of beauty:

> The sunset I viewed from the crest of the (Haina) hill captured an exceptional mood. At my feet was the endless, luminously green expanse of bush enclosed by the hazy blue Kwebe Range; rising above them were stacked masses of cloud glowing red and gold in the thundery sky. It was a magnificent view (Passarge 1904a: 225).

All through his two and a half years in southern Africa Passarge corresponded formally and informally with Von Richthofen in Berlin. His letters from Kimberley and later from camps along the Boteti River and Lake Ngami as well as from Andara on the Okavango were published in *Verhandlungen der Geselschaft für Erdkunde zu Berlin* and summarized in *The Geographical Journal.*

By the end of 1898 it was clear that West Charterland's dream of minerals in Ngamiland was a mirage. Passarge was released from his contract and he left South Africa in January 1899 with an ample bank balance and stacks of field notes. Home in Germany he became a popular speaker at the meetings of local geographical societies on the topic of southern Africa, where war clouds were looming. However, his notes on the subcontinent still had to be written up and interpreted.

The three southern African monographs

Passarge soon realized that writing on southern Africa, where a scientific discourse was emerging, presented different problems than writing about the unknown Cameroon. By 1896 the Witwatersrand was already one of the world's premier mining regions; medium-scale maps, admittedly coarse in resolution, were available for areas such as the Western Cape and the Natal-Transvaal borderland; in Cape Town a Meteorological Commission was collecting rainfall data and ethnographers had at least a general idea of the subcontinent's human diversity. What was lacking were frameworks within which such knowledge could be conceptualized: a reliable stratigraphic column, a functional theory of the subcontinent's climate, and an explanation of the origin of the subcontinent's ethnic contrasts. The researcher's task was not only to add detail but also to develop new frameworks and assess existing ones. The pioneer, the person who first ventured into a new field, is especially vulnerable because later researchers and new evidence might necessitate new conjectures and the rejection of the earlier somewhat explanatory frameworks. This was eventually Passarge's destiny and he didn't suffer it gladly.

Passarge's notes on southern Africa proved an inexhaustible source of empirical information and among his 45 publications on the subcontinent, three substantial monographs became benchmarks in the region's geographical literature: *Die Kalahari* (1904a), a physiographic reconnaissance in the style of Von Richthofen's classic on China, *Die Büschmanner der Kalahari* ("The Bushmen of the Kalahari", 1907) which displayed the insights of medical training in observing the subcontinent's oldest ethnic group; and *Südafrika: Eine Landes-, Volks- und Wirtschaftskunde* ("South Africa: A Study of its Land, Peoples and Economy", 1908), the first full-length regional geography of the subcontinent written by a professional geographer. It is worthwhile to look at them separately.

Die Kalahari, which was also Passarge's dissertation for *Habilitatis* at Berlin University under the supervision of Von Richthofen, was a work of Teutonic thoroughness: 822

pages, 38 chapters and an atlas of 21 folded maps and diagrams. The introductory five chapters contextualized the Kalahari region within the framework of southern Africa's relief, geology and climate. Chapters 6 to 29, the bulk of the book, were exploration reports: detailed descriptions of the specific inselbergs, watercourses and pans where Passarge searched for minerals, places with names like Kwebe, Tsodilo, Monekau, Epukiro and Gobabis, each mapped according to access routes, morphology and geology. In the final nine chapters Passarge was once again the scientific geographer who interpreted his geomorphological and stratigraphical observations according to cause, origin, history and consequences.

First among the problems Passarge addressed was the fossilized drainage systems of Ngamiland. He regarded them as evidence for a progressive desiccation of the subcontinent which commenced in the Mesozoic and was still in progress. Lake Ngami, he thought, dried out only ten years earlier. Many of the pans were biologically maintained: first ants, termites and small mammals burrowed underneath the moist depressions, then vast herds of game trampled and wallowed in the soft soils, mud collected on their hoofs and skins and provided the dried clays and silts which were eventually removed by deflation. Passarge also recognized the Kalahari sediments as a separate stratigraphic sequence and explained the silcrete and calcrete duricrusts as precipitates which formed under progressive desiccation. The quintessential African landscape, however, was the combination of inselbergs and basal plains of the high interior. Ngamiland, Passarge thought, held the key to their origin. The inselbergs were invariably built of resistant granite while the thinly veneered basal plains were underlain by softer gneiss and schist. Only an agent acting horizontally, randomly and sometimes against the slope could be responsible for such a large-scale abrasion feature. Passarge sought the cause in fierce winds which supposedly swept the extreme interior deserts of the African supercontinent during Triassic and Jurassic times when a dry Indian Ocean basin provided a continuous linkage eastwards to India and Australia.

Passarge's observations of the widely scattered inhabitants of the northern Kalahari were summarized in a number of ethnographic reviews and a monograph of 150 pages on the San, in the terminology of the time, *Die Büschmanner der Kalahari* (1907). At the end of the nineteenth century the Kalahari San still lived beyond the formal colonial frontier and Passarge presented a sympathetic picture of their untouched existence. With his medical attentiveness he could not miss their physical resemblance to the surrounding Bantu. They were gatherers rather than hunters, enmeshed in a bartering network with their more advanced neighbours but spoke nine different languages. Passarge also noted the San's ability to organize, a characteristic previous observers had missed. In semi-arid regions like the Gobabis ridge across the border in German South West Africa, where waterholes survived for seven years after a good rainy season, the San congregated in groups up to 500 individuals and a particular chief could marshal 100 of them for a hunting drive.

Whereas *Die Kalahari* and *Die Bushmänner* were based on empirical research and addressed Passarge's peers, *Südafrika* (1908a), the third monograph, was a professor's extension lecture for the informed public. The book was propitious. The South African War still evoked memories; the British had their "Khaki" election in 1901; Chinese labour on the Witwatersrand was a hot item in British politics; and the Germans, perturbed by their inability to quell Nama resistance in their South West African colony, had a "Hottentot" election of their own in 1907.

Following Ritter and Von Richthofen, geographers regarded area studies, or what was eventually called regional geography, as their distinctive contribution to science but were uncertain about its structuring. In Germany the controversy between the proponents of *Länderkunde*, the study of politically demarcated areas with formal boundaries, and those of *Landschaftskunde*, the study of areas identified on account of natural features, was just heating up. Passarge used both for *Südafrika*. He regarded "South Africa" as the subcontinent south of the watersheds of the Zambezi, Okavango and Cunene Rivers with the Congo. In a text of 341 pages, however, this vast area could only be coarsely covered. The section on physical geography was ordered according to the *Landschaftskunde* model: first a sequence of chapters on relief, drainage, climate, geology, vegetation and animal life and followed by a detailed synthesis of fourteen specific natural landscapes. The chapters on human geography were more eclectic: first a section on the habitability of the subcontinent according to its natural environment; then chapters on population distribution, physical and mental characteristics of the different ethnic groups, languages, prehistorical and contemporary indigenous cultures; and finally, brief *Länderkunden* of the twelve colonial territories, emphasizing economic activities stimulated by the imperial powers.

Passarge, notwithstanding *Südafrika's* coarse resolution and the eight years lag since leaving the country, remained well informed of themes in the local scientific discourse. Earth scientists, for instance, had long been speculating on the origins of the Great Escarpment and by 1900 consensus was that the high east face of the Drakensberg was a fault phenomenon, the so-called Quathlamba Fault. Unacquainted with the eastern part of the subcontinent, Passarge accepted this verdict on circumstantial evidence. The mountains of the southern Cape, he observed in *Südafrika* (1908), owed their prominence partly to faulting on their coastward-facing slopes and the same might be true of the Drakensberg since the Basutoland basalts suggested a fractured crust. He reiterated his belief in the progressive desiccation of the subcontinent and introduced the notion of a "Pluvial", a postulated rainy epoch in the subtropics which coincided with the last glacial episode of the temperate latitudes. During the Pluvial the whole northern Kalahari was supposed to have been an extensive marsh which, apart from the Okavango Swamp, had since evaporated.

Passarge's human geography, in contrast to his physical geography, was already the perspective of an outsider 8 000 km away and displayed little awareness of local political discourses. At the turn of the nineteenth century most Europeans explained cultural differences as a matter of race and accepted that which they regarded as black backwardness proven by science. Passarge took an extreme position in supporting this fallacy and casted the relationships among indigenous ethnic groups in a mould of rigid social Darwinism. Southern African ethnic groups, he claimed, struggled among themselves for resources: the stronger groups seized healthy arable areas of preference (*Vorzugsgebieten*) like the Zulu in Natal, while weak groups like the San in the Kalahari were driven into harsh areas of retreat (*Rückzugsgebieten*). White-black political discourse was presented as stark racial confrontation. Blacks were again dismissed as frivolous, unpredictable and thoughtless, the Khoi and San were even worse, while the rise of separatist black churches was presented as a symptom of hatred towards whites. Therefore, "every act of humanity towards blacks is at the same time cruelty towards whites" (Passarge 1908a: 340). The answer: whites should treat blacks as "an inferior race". Passarge also missed the finer nuances of white political discourse: the English vision of a Greater South Africa, the Afrikaner yearning for unity across colonial boundaries and the search of both groups for reconciliation. But the South African War left him with no illusions about Afrikaners. "A people of whom one third are bribable louts and traitors, must perish" (Passarge 1908a: 205).

The aftermath: Getting established in an academic career

The impact of Passarge's southern African publications struck at two levels. Personally they provided a base for an academic career. Reviews of all three monographs were overwhelmingly positive: both German and British reviewers were impressed by the empirical thoroughness of *Die Kalahari* and at least one German reviewer was persuaded by Passarge's arguments for the desiccation of the subcontinent (Hahn 1908) but the *Geographical Journal* (1908) found *Südafrika's* summary of black-white relations "jarring". In the process Passarge accumulated all the prerequisites for an academic geographer: as an international expert on arid geomorphology, *the* German authority on southern Africa and a respected consultant on German colonial problems. In 1900, Passarge's intervention in a contract between a German and a British firm on the construction of a railway line across the north of German South West Africa impelled the German Government to cancel the contract.[3] Passarge also found time to visit Venezuela to assess an estate for a Cologne buyer and often commented on development problems in Cameroon.

Competition for positions in geography at Wilhelminian universities was, however, intense and Passarge struggled for three years as *Privatdozent* in Berlin before being called to a well-established chair at Breslau University (today Wrocław, Poland) where most of *Südafrika* was written. In 1908 he moved to Hamburg to run the Geographical Seminar, a division of the German Colonial Ministry's newly established Colonial Institute. There he was tasked with training colonial officers and doing research on overseas territories.

Passarge, meticulous researcher, African expert and ardent nationalist, was just the person for the job.

Regarding scholarly discourse Passarge did not shy away from polemics. When expert opinion disagreed with some of his explanations he responded wrathfully. The veteran ethnographer Gustav Fritsch (1835-1927), for whom the San were cultural isolates, "unhappy children of the moment", faulted Passarge's portrayal of them as a group with organizational ability, outside contacts and a history. Of course, Passarge rejoined, all humans were social and the San were not different from other people (Wilmsen 1997).

A debate developed on southern African landforms and it spread from concepts to persons. When William Morris Davis, the leading Anglophone geomorphologist of the time, read Passarge's papers on inselbergs, he hit on the latter's basal plains as the explanation for the final, "old" stage of his arid geographical cycle (Davis 1905a,b). Recasting the process as "levelling without base levelling", Davis used Passarge's explanation in the paper he presented at the Johannesburg session of the joint S_2A_3–British Association conference of 1905. But during the conference excursion when he saw the gently undulating Highveld surface, he had second thoughts and kept open the option that the southern African interior plateau might also be an uplifted peneplain.

Albrecht Penck, another participant in the excursion and soon to be Von Richthofen's successor in Berlin, was sympathetic to the idea of an uplifted basal plain, and he saw the Great Escarpment as the result of upwarping so ascribing the towering profile of the Drakensberg to differential erosion (Penck 1908). Passarge, forgetting of his own flimsy reasoning, exploded when he read Penck's papers. "Penck's observations from the railway track", he sneered, "can only be of limited value" (Passarge 1908b: 141). Two years later, when a student of Penck offered a more satisfactory explanation for the peculiar limestone pans of the western Kalahari, Passarge declared Penck enemy for life. Eventually, Davis was also accused of being better at meditation than at fieldwork (Passarge 1951).

By the 1930s nobody still took an aeolian origin of inselberg landscapes seriously, while Penck was clearly on the right track with his explanations of both the Great Escarpment and limestone pans (Werther 1935). His ideas on the Drakensberg led through Obst & Kayser's (1949) *"Piedmenttreppen"* and LC King's pediplains (1962) to Partridge & Maud's (1987) concept of denudation surfaces.

Passarge opened the South African geographical discourse and gave it an international character but he paid the price for being the first to formulate concepts which he never revised and applied to a subcontinent to which he never returned.

Hamburg: The polemical years

With a population of 950 000 Hamburg, where Passarge worked until his retirement in 1935, was Germany's second city and leading port, with a worldwide oceanic foreland and the productive Elbe basin as hinterland. The First World War brought ruin: from 1914 to 1918 the port was severed from its foreland by a British naval blockade, while after the war Germany lost its colonies and the Elbe hinterland was disrupted by the borders of the new Poland and Czechoslovakia. Like other German ports the city experienced a collapse of law and order after the Armistice of 11 November 1918 because sailors were the bellwethers among the mutineers in the armed forces (Watt 1968). A fragile order returned by the middle of 1919 and the City Council used the lull to integrate the city's three centres of higher education, among them the now functionless Colonial Institute, into Hamburg University. Both in the city and in the university senate, socialist sentiments prevailed but Passarge's chair and the Geographical Seminar, now simply a geography department, were transferred to the new institution.

Like many of his conservative compatriots Passarge found the events of 1918 and 1919 inexplicably terrible and he fell into deep pessimism. Germans tried to explain their malaise in at least two ways. On the one hand, the collapse was seen as a manifestation of the downward limb of an impersonal cycle of cultural growth and decline experienced by the West as a whole but, because of the pressures of war, especially severe in Germany. This view was classically articulated in Oswald Spengler's *Untergang des Abendlandes.* By contrast, other thinkers – and they were the majority – looked for social pathologies closer to home: the unwillingness of the German people to unite in the face of crisis, the lack of discipline among certain segments of society and blatant treason by socialists and communists, the mythical stab in the back. Passarge accepted the second version and circuitously weaved it into his theory of geography.

Early in his Hamburg years Passarge had already given thought to an improved structuring of area studies and eventually he chose the concept of *Landschaft* as his key. After all *Landschaft,* in contrast to *Land,* was nature-given. By 1920 the debate between the supporters of *Landschaftskunde* and *Länderkunde* was approaching a climax but the former were losing because they disagreed among themselves on the exact meaning of *Landschaft.*[4] Passarge searched for a clear definition of the concept and its rational application in research. He interpreted *Landschaft* as a small bounded tract with a unique combination of empirically recognizable features and labelled by a neutral name revealing nothing of its origin or function. *Landschaften* were supposed to be the building blocks of geography, the equivalent of species in biology and elements in chemistry, and Passarge suggested a hierarchy along which individual *Landschaften* could either be subdivided or aggregated.[5] The researcher was supposed to commence with the particular part of the Earth's surface, teas out its landscapes and landscape hierarchies, choose the level in the hierarchy at which the research should be done and only then start to look for causes and explanations. The procedure was ponderous and difficult to apply but for Passarge *Landschaftskunde* became a distinct sub-discipline of geography and he devoted volumes of formidable size and complexity to its practice (Passarge 1919, 1921, 1923).

Passarge's *Landschaftskunde* had a bizarre twist. He recognized that human intervention could modify natural landscapes into agricultural and urban landscapes. All landscapes also determined the character of their inhabitants. The consequences of such controls were spelt out in two little known books, the first innocuously named *Beobachtungen über Mensch und Tier* ("Observations on humans and animals", 1922) and the more formal *Grundzüge der Gesetzmässigen Characterentwicklung der Völker* ("Outline of the Law of the Development of the Character of Peoples", 1925). Everywhere, Passarge asserted, peoples (*Völker*) were involved in a struggle for existence and the nature of this struggle was closely tied to the landscape where a community lived. In natural and agricultural landscapes people fought with nature and learned to be hard-working, steadfast and self-reliant. But in cities with their protective walls, people had no need to defend themselves and they split into groups who struggled with each other. Some peoples degenerated into docile and servile individuals, quick-witted but also intentionally deceitful, crafty and opportunistic. Regarded as an urban class, Passarge called them Sarts, a Russian term for the Jewish and Armenian small traders of the oasis towns of Central Asia. "Sartification", Passarge declared, was a logical consequence of his law of the development of character and the machine cities of the industrialized twentieth century, with their restless proletariats, were exceptionally vulnerable.[6] The presence of Sarts in German cities was the country's ruin. For the healing of the national culture there was only one cure: the Sarts were a cancer to be removed from the body of society (Passarge 1925). *Grundzüge* went one step further than anything Passarge had written before.

Because of its convoluted nomenclature, stark environmental determinism and doubtful consequences, Passarge's *Landschaftskunde* had few followers. "It is difficult to understand how (this) arid method can be advocated by a man so experienced in exploration and so expert in explanation as Passarge", was *Geographical Review*'s (1919) comment on *Grundlage der Landschaftskunde* (1919), while the *Geographical Journal* (1930) found the machine city and degenerated Sarts "startling excursions into other realms of science." But Passarge resolutely defended his fallacies and became entangled in endless polemics, especially with proponents of *Länderkunde* such as Alfred Hettner of Heidelberg and Alfred Philippson of Bonn,[7] both of whom had Jewish family connections and, of course, Albrecht Penck, who did not share his ideas of national revival. In 1927 at Hamburg University Passarge harassed his Jewish colleagues with such venom that the Rector had to warn him formally that there were limits to academic freedom. By the 1930s solid research papers in standard journals from Passarge's pen dwindled and his interest in geomorphology all but disappeared. The leftovers were extensions of his *Landschaftskunde* volumes of the 1920s and reviews resurrected from visits to Latin America and North Africa many years before.

On 30 January 1933 Adolf Hitler was sworn in as German chancellor. This left Passarge (1937: 85) enraptured:

> When the banner of national socialism unfolded itself triumphantly over Hamburg University, the Geographical Seminar did not have to change. Here were no Jewish-Marxist lectures and assistants, no Jewish-Marxist students who made themselves unpleasant. Here was no portrait of Ebert to be removed. Here was no change in the character of the lectures. Inwardly the Geographical Seminar was already attuned to the new age.

The Nazi's were appreciative. In May 1933 Passarge was awarded with the position of *Reichsobmann* for geography in the Nazi teachers' union. In November 1933, after an arrogant letter to the French secretary general of the International Geographical Union, Passarge was dismissed as too combative. He became more embittered and irascible than ever.

Ironically, Passarge had a second chance. During the Second World War German geographers were drafted into government to assist in the planning of conquered Eastern Europe, only to sneak back to their chairs in 1945. But the university courses of the 1930s had to be "denazified": in geography *Raumordnung* and *Geopolitik* were out and *Landschaftskunde*, stripped of machine cities and Sarts, was in. As a pensioner, Passarge was untainted by pogroms and labour conscription in the East and his ideas on the identification, subdivision and aggregation of *Landschaften* showed a way to a non-political physical geography. In 1953 he received the prestigious Ritter Medal from the *Gesellschaft für Erdkunde zu Berlin* and in 1956 his former employer, Hamburg University, awarded him an honorary doctorate (Louis 1959; Kanter 1960). It was left to a younger and more critical generation of German geographers to expose the dark side of his geography (Sandner 1988, 1992, 1998).

The ambivalent heritage

What is Passarge's significance as a South African founding geographer? Two perspectives are possible. The biographer may view his person, career and publications as a whole because, as Sandner (1990) remarked, *Landschaftskunde* and the law of character development were bricks of the same building. There is a danger, however, that the biographer may overemphasize those parts of the structure which were built later and are therefore closer to his own time and problems. The alternative is to interpret Passarge's career as a trajectory which passed through different places at different times and involved him in different discourses.

Passarge's immediate successors were the young German geographers Leonard Schulze, Fritz Jaeger and Leo Waibel who visited the subcontinent before the First World War and Erich Obst and Kurt Kayser who came post-war. They all maintained Passarge's empirical approach, retraced his themes – aridification, pans, the Great Escarpment – in finer resolution and corrected some of his errors. However, the disruption of war and superinflation in Germany prevented them from completing their projects and many of their results went unpublished. "The Versailles Treaty", Jaeger (1925: 289) wrote in bitterness, "not only robbed us of our colonies but drove us from the field of African research." No wonder that Jaeger, Waibel and Obst also turned to the political right in their later years.

Although South African scientists sometimes disagreed with Passarge, they respected his pioneering work. Alex du Toit (1924) recommended *Die Kalahari* as a model for an exploratory study of any explained part of the subcontinent and he appropriated Passarge's fivefold subdivision of the Kalahari sediments in his *Geology of South Africa* (1925) which remained the standard work on the geology of southern Africa until the 1960s. In Botswana, Passarge's stratigraphy was tolerated into the 1980s but increasingly accurate age determinations have suggested that scattered sediments of similar lithology may date from totally different times of the Tertiary (Shaw 1993). A revision is urgently necessary. Meanwhile, one of the longitudinal depressions running out to the Makgadikgadi Depression from the south-west carries Passarge's name (Thomas & Shaw 1991).

Die Buschmänner had an even longer life. The University of Cape Town anthropologist Isaac Schapera (1905-2003) depended heavily on Passarge's ethnographies for his *The Khoisan Peoples of South Africa* (1930) which was the authoritative work on the two ethnic groups until the 1960s. During the 1980s a debate on the prehistory of the San flared up among anthropologists and smouldered into the 2000s. A group of American anthropologists, since known as the "isolationists", studied San communities roaming in the Namibia-Botswana borderland twenty years earlier and idealized their lifestyle as "the most successful culture mankind has ever known" (Thomas 2006: 6). They were accosted by the "revisionists", both Europeans and Americans, who argue that the San, even before colonial times, had been and still are as exposed to change, even stress, as any other southern African community and invoked Passarge as authority (Barnard 1992). This debate, Wilmsen (1997) has suggested, would have been unnecessary if Americans took the trouble to read German: in 1980 only three copies of *Die Bucshmänner* were available in all of America's libraries.

As Passarge passed on to the Hamburg sector of his trajectory, his connections with Southern Africa rubbed thin. Of his 216 publications since 1919 only twelve related in some way to the subcontinent. Geography departments appeared in South African universities from the 1920s but there is no record that he solicited contact with any of them. Of the ten doctorates in geography awarded by German universities during the interwar years, the Geographical Seminar handled only two (Dietzel 1940), one of which was that of Gerrie Scheepers (1938) who chaired the Department of Geography at Pretoria University from 1950 to 1974.[8] Although Passarge dropped out of the southern African discourse, the problems he had recognized challenged the founders who followed. Schwarz of Rhodes University College became obsessed with the Kalahari, Serton of Stellenbosch critically discussed *Landschaftskunde* with his graduate students and Wellington's *Southern Africa* (1954, 1955) was an intentional successor to *Südafrika*.

Endnotes

1 Chapter 3 draws heavily on two unpublished sources at the library of the Geographical Institute of Hamburg University. The first is a typed autobiography of Passarge with the title **Aus achtzig Jahren: Ein Selbstbiographie** ("From eighty years: a self-written biography", 1951). Some of the author's remarks in the text, e.g. on pages 166 and 310, suggest that most of the document was compiled during the early years of the Second World War and acquired by the Institute in 1951. The second is Sandner & Rössler's **Schriftenverzeichnis von Siegfried Passarge** ("Index to Siegfried Passarge's scripts", 1987), a bibliography listing 337 of Passarge's publications and manuscripts: books, papers in journals, book reviews, maps, commentaries and unpublished essays.

2 Direct quotes are translated into English but German is maintained for book titles and geographical terms such as **Land, Länder, Landschaft** and **Landschaftskunde** for which English equivalents are unsatisfactory.

3 Late in 1899 the De Beers Company of Cecil Rhodes took over the mineral rights of British West Charterland in the north-western Bechuanaland and negotiated a deal with the South West Africa Company, a German firm, to mine diamonds which might be discovered in the Etosha Pan area. Should such a venture become viable, De Beers undertook to finance a railway line from a port in the south of Angola via Etosha and Ngamiland to Rhodesia, a route which would have been of scant benefit to the German colony. Through his contacts with British West Charterland, Passarge got wind of the negotiations and criticized the "unpatriotic" directors of the South West Africa Company to such an extent that the German Government vetoed the deal (Passarge 1951).

4 The controversial history of **Landschaft** derives from the different meanings ascribed to the word in the German language. In Old German **Landschaft** was the equivalent of the Latin **regio**: a demarcated tract sufficiently small to be observed from one viewpoint and held by a lord. **Land** was a bounded area too large to be observed in one sweep and held by a noble of higher rank, e.g. a duke. When paintings of scenery became popular through the work of the Dutch School of the seventeenth century, **Landschaft** acquired a second and slightly different meaning: an area visually observed and aesthetically assessed. In this sense the word was transferred to English and its visual connotation emphasized by parallel terms such as a "seascape" and "cityscape". Most German geographers – Passarge among them – regarded **Landschaft** as the equivalent of "region" but others, such as the German Otto Schlüter (1872-1952) and the influential American CO Sauer (1889-1975) restricted it to the visually observable elements of the Earth's surface. (See Hartshorne 1939; Beck 1973; James & Martin 1981).

5 While **Landschaft** was the basic areal unit, Passarge also provided for areal studies at other scales. Should a larger scale and finer resolution be necessary, a **Landschaft** could first be subdivided into **Teillandschaften** (landscape sections) and then into **Landschaftsteilen** (landscape parts). At smaller scales **Landschaften** could be aggregated into region-sized **Landschaftgebieten** (landscape areas) and global **Landschaftsgürtel** (landscape zones) (West 1990). The difficulty of applying this complicated scheme in the days before computerized geographical information technology is obvious.

6 Passarge's **Beobachtungen über Mensch und Tier,** published in Gothic lettering and dedicated to the nationalistic-minded German youth, was obviously aimed at the **Wandervögel** (literally: migratory birds), a movement among the conservative middle-class youth of the late nineteenth and early twentieth centuries who went marching, singing and camping in the countryside (Fulbrook 1990). On the surface these youth were clean-cut, muscular, supportive and high-minded, but they were also intensely nationalistic and suspicious of the megacity with its "degenerate mass culture and greedy capitalists", sentiments which were a potent seedbed for national socialism. Passarge, whose love

for nature was genuine and who had the grit to overcome a harsh environment, was vulnerable to these views.

7 Sandner (1990) illustrates Passarge's polemical style with excerpts from an altercation he had with Alfred Philippson in the columns of the *Geographische Zeitschrift* (1929). In a review of a book by Passarge on southern Italy, Philippson, a specialist on Mediterranean countries, pointed out numerous factual errors: carelessness (*Leichfertigkeit*) was his verdict. In his rejoinder Passarge shifted the argument from the book to the person: "Because of the brevity (of the text) some unclarities seemed to have slipped through but they obviously served as a basis for an emotional personal attack." In a follow-up letter to the *Geographische Zertschrift* he accused Philippson of a deep personal hatred towards him and sent copies to both his own Faculty at Hamburg University and the Prussian Ministry of Culture. No, Philippson retorted, what is at stake, is the integrity of the discipline.

8 Scheepers' dissertation was entitled *Maisanbau und Klima im Sogenannten Maisdreieck der Südafrikanischem Union* ("Maize Cultivation and Climate in the so-called maize triangle of the Union of South Africa, 1938"). The research had no obvious links with either Passarge's earlier southern African monographs or his *Landschaftskunde*.

CHAPTER

4

Schwarz of the Kalahari

Like Passarge, Ernest Hubert Lewis Schwarz (1873-1928) qualified as a geologist but unlike Passarge he remained one and built his career in South Africa. Between 1890 and 1920, especially in English-speaking countries, the boundaries between geology and geography were permeable. Domains overlapped, concepts were shared, field geologists commenced their reports with an overview of landforms, and innovative geographers like Davis and Penck were in demand at geology conferences. Ernest Schwarz was comfortable in this borderland. Shortly after joining Rhodes University College in 1905 he published papers on stream piracy (Schwarz 1906) and uplifted marine benches (Schwarz 1907), studies which today would be regarded as geography. Between 1918 and 1928 he was responsible for the courses in physical geography and economic geography at Rhodes, served for eight years on the University of South Africa's examination committee for geography and, according to a colleague, never forgot that humans were the most interesting of natural phenomena (ARL 1929). Geographers, in turn, accepted him: "...a distinguished geographer of wide repute and intrepid opinion" were the words with which he was welcomed as president of the South African Geographical Society (Anonymous 1922: 4).[1]

The trajectory of Schwarz's career reached its climax with the Kalahari Scheme of 1918. His Kalahari was an imagined place, conceived from afar in an encounter between the discipline of geography and the spectre of drought. The scheme, its underlying assumptions, the driving motives and the perceptions of the southern African environment uncovered by the ensuing controversy, occupy a central position in this chapter. But earlier nodes in Schwarz's trajectory are also important: Late Victorian England where he received his education and acquired a love for writing and the theatre; the Cape Folded Belt where he won his spurs as an Earth scientist and was distressed by the ravages of drought; and Grahamstown where his work as a university teacher widened his horizons and inspired him with bold visions.

London, Camborne and the Cape Colony: The empirical years

Ernest Hubert Lewis Schwarz was born on 27 February 1873 in Basinhall Street in the London suburb of Sydenham, the son of FM Schwarz, a merchant in the South American trade. Situated 10 km south-east of the City, Sydenham in the 1870s was a spacious semi-rural residential area favoured by upper-middle class families. For his final years at school the young Ernest travelled across the Thames to the venerable Classical School of Westminister where the English public school tradition of Greek and Latin still held sway. Once a year since the sixteenth century the boys staged a play in Latin. Schwarz was keen on dramatic art all his life, quoted Greek poetry with appreciation and had a skilful and facile pen (ARL 1929). But it was science which caught his interest.

Education in the natural sciences faced many obstacles in Victorian England. In Germany the academic freedom of universities following the Von Humboldt ideal promoted intellectual entrepeneurship, they competed for students and professors, and new disciplines and departments evolved throughout the nineteenth century. Two universities, Oxford and Cambridge, dominated the scene in England but they were muzzled by ancient traditions and looked down on the applied sciences as skills for tradesmen. Ironically, industry and technology made the United Kingdom "the workshop of the world": a country prosperous, peaceful and relatively stable. By the second half of the century some leaders of industry had become frustrated by this indifference to scientific training and joined hands to establish private institutions for providing specialized education in the pure and applied sciences. One of these was the Royal College of Science, assembled in 1881 by Thomas Huxley from three smaller predecessors and sited at South Kensington, the intellectual heart of London also north of the Thames (Gray 1997).

Ernest Schwarz entered the College in 1891, was soon fascinated by geology and obtained his associateship with distinction in palaeontology. But a geologist had to earn his bread. Schwarz concluded his education at Camborne School of Mines 25 km from where Cornwall, the Ancient source of tin, tapers off into the Atlantic Ocean. Camborne's courses specialized in advanced mineralogy, mining geology and metallurgical engineering which gave Schwarz a thorough professional edge. But there was another side to the coin. At that stage neither the College nor the School required a thesis through which a student learned to synchronize theories with observations, a weakness that stalked Schwarz throughout his career.

In 1895 South Africa was the one place in the world where an aspiring 22-year-old geologist wanted to work. Schwarz embarked for South Africa to study the burgeoning gold-mining industry and obtained a job as editor of a Johannesburg monthly, *The Scientific African* (Schwarz DM 1929). As with Passarge, the Witwatersrand did not appeal to Schwartz and by the middle of 1896 he moved to Cape Town to become second field geologist at the Geological Commission, Cape of Good Hope, then in its first year of activity (EDM 1968). The Commission was one of the earliest and most successful outcomes of the wave of scientific institutionalization sweeping South Africa in the 1890s. Initially advocated by the Philosophical Society of South Africa, the Cape premier Cecil John Rhodes sat on the parliamentary committee which recommended the Commission's establishment and the institution's goal was purely and simply scientific: the systematic exploration of the geology of the Cape Colony and the summary of survey results on reliable maps (Anhaeusser 1997). The Commission also valued academic connections. The part-time director, GS Corstophine, was simultaneously professor of geology at the South African College (SAC), while the first field geologist, AW Rogers (1872-1945), who was to leave deep marks in the South African Earth sciences, lectured at SACs from time to time. Rogers and Schwarz soon settled into a working routine: together they went into the field for periods of six to ten weeks, thrashed out problems of interpretation on site and returned to their offices at the South African Museum adjacent to SAC, to compile reports and maps.

The Commission's aim was to extend its surreys systematically eastward from the south-western Cape until the whole of the Cape Fold Belt was mapped. Generally, Rogers and Schwarz stuck to and completed this task although they were briefly deflected eastwards into the Transkei and Griqualand West by the activities of Boer commando's during 1901. By this time South African geologists had hammered out a usable stratigraphic column for the subcontinent as a whole and the major rock units of the south of the Colony found their niches but the identity of many specific outcrops was uncertain and almost nothing was known of their structure.

Rogers and Schwarz started close to Cape Town in 1896 and followed the westward mountain front overlooking the Boland and Swartland regions northwards. They recognized the open anticlinal nature of the Wemmershoek and Du Toit's Mountains, crossed the Cederberg, noted the grotesque weathering forms and erosional benches along its eastern outcrops and later identified the shale horizon which divides the Table Mountain sandstones in two separate sequences (Rogers & Schwarz 1901). In 1897 the Langeberg came under review. The two geologists investigated the Worcester Fault for 150 km along its southern slopes between Tulbagh and Swellendam, established its age as Early Cretaceous and teased out the enormous displacement of 3 750 metres between Worcester and Robertson. In 1898 they traced faulted environments further east, namely the discontinuous exposures of the Worcester Fault until it finally disappeared under the sea at Glentana and, to the north, the bold and massive Cango Fault along the southern slopes of the Swartberg. After the Transkei interlude and the Peace of Vereeniging Schwarz, now working on his own, returned to the eastern Folded Belt and during 1903 and 1904 mapped the complex structures of the Baviaans and Langkloof areas in the hinterland of Port Elizabeth (Schwarz 1904, 1905). He was struck by the more intense folding of the eastern belt, the lateral unevenness of the Baviaanskloof

Fault, its "imbricate fault-pits" filled with younger Cretaceous conglomerates and the high-level terraces truncating the slopes of the Baviaanskloof and Kouga Mountains.

Schwarz's empirical work between 1896 and 1905 in the Cape Fold Belt was his most enduring contribution to South African Earth sciences. The assignments were specific, he had to keep his eyes on the ground and his conclusions derived directly from his observations. Schwarz took pride in doing fieldwork during times of rinderpest, war and drought. He covered long distances by wagon along remote tracks, shifted his camp often and for days traced rock contacts for days on foot over craggy terrain. During these exercises he learned to observe more than just geological formations: rock paintings in the Cederberg, furrows and irrigated fields in the Little Karoo and the Willowmore cedar which carries his name (*Widdringtonia schwarzii*) (Lückhoff 1963). The austere beauty of these lonely mountains appealed to him and he listened to its people: Klaas Trompeter, his diminutive wagon "boy" at the camp fire (Schwarz DM 1928) and, in the deep valleys of the Kouga, farmers with tales of drought and toil. White poverty was rampant in the region and Schwarz wished that he could assist these people in their distress.

Grahamstown: Distant views from the Suurberg

The year 1904 brought fundamental change to Schwarz's life. On 20 April he married Daisy Murray Bowne in Cape Town: attractive, intelligent and a staunch ally in the years ahead. In the corridors of the South African Museum there were rumours of discord between Schwarz and his colleague Rogers who succeeded Corstophine as director of the Cape Geological Commission in 1902. Temperamentally they were opposites. While Schwarz was "always attracted to the speculative side" (PL 1929: 3), Rogers was an arch empiricist. "To be a good geologist", he told his students (Brown 1977: 341), "you must remember three fundamentals which are observe, observe, observe." Both, however, were decent persons and their differences never flared up into unpleasantness. Schwarz, then 32 years old, also sought a more settled life and in 1905 accepted a chair in geology at the recently established Rhodes University College.

Grahamstown was established in 1812 as the headquarters of British garrisons on the Eastern Cape frontier. In 1828 the military encampment became the main administrative centre for the British settlers in the Albany district and, as the colonial border shifted from the Fish to the Kei Rivers in 1847, assumed cultural functions for the Eastern Cape as a whole. An Anglican bishop was consecrated in 1853, exalting the town to the status of "First City", and other institutions followed: the Albany Museum in 1855, a branch of the Cape Supreme Court in 1865, an art gallery and eventually a range of church schools of different denominations, all cast in the mould of the English public school. However, when diamonds were discovered in Griqualand West in 1869, the railways from the two Eastern Cape ports bypassed the uncomfortably situated Grahamstown, so leaving the city, in the words of one commentator, with "a healty environment, physically, socially and morally and accessible from all parts of the Eastern Province" (*The Rhodian* 1906: 2-12).

By the late nineteenth century the fame of the city's schools had already stirred ambitions for the establishment of a centre of tertiary education. From 1876 the Anglican Saint Andrew's College prepared postmatriculation students for some of the degree examinations of the University of the Cape of Good Hope and had by the turn of the century a staff of four professors who had the training of surveyors as their most important specialization. The conclusion of the South African War in 1902 brought optimism, confidence and high expectations to the English-speaking Eastern Cape. In December of that year the editor of the *Grahamstown Journal* launched a campaign to establish a non-denominational university college in the city but soon discovered that the cost of such an institution was more than the Eastern Cape could muster. The initiators turned to the Rhodes Trust, committed themselves to name the college in Rhodes' honour, sounded the imperial trumpets and after some bargaining secured a grant of £50 000. "Rhodes University College", reported the Cape Town secretary of the Trust to his London superiors (Currey 1970:12), "is designed ... to extend and strengthen the Imperial Idea in South Africa, where so far the only decent University education to be had, is at Stellenbosch, under influences notoriously anti-Imperialist." The university college came into being on 1 July 1904 when the four professors and 39 undergraduate students moved from St. Andrew's College to the Old Drostdy at the top of High Street. On 1 May 1905 two additional professors assumed duty: the zoologist JE Duerden and the geologist EHL Schwarz. So cash-strapped was the university college that up to 1910 the two gentlemen had to share their time and salaries on a 50-50 basis with the next-door Albany Museum.

Ernest and Daisy Schwarz settled in a house in Constitution Street along the fashionable West Hill just beyond the St. Andrew's grounds. With a population of 14 000, more than half of them detribalized Africans, Grahamstown was, in South African terms, a large and sophisticated town. The streets, long, straight and dusty, were sufficiently wide to allow a team of oxen to turn around and they were sprinkled with water on windy days. Residence styles were eclectic but harmonious and the city, dominated by the towering cathedral in High Street, had dignity. Social life was provincially English with Founder's Day dinners, amateur theatricals, soirées, tennis teas and popular lectures on scientific subjects (Collier 1961). The Schwarzs remained childless but they flourished in the convivial atmosphere. Ernest, "tall, with a small beard and gold-rimmed glasses" (Du Toit 1945: ix), was a lively raconteur with a touch of imagination. During his years at the Geological Commission he worked on a play and shortly after settling in Grahamstown he ventured his hand at a novel which was published under a pseudonym. But beneath the surface he was shy and sensitive. "Disappointment", a colleague (ARL

1929: 4) observed, "caused his genial and expansive nature to withdraw itself so as to chill for a time even the warmest of friendships."

Schwarz's sociability made him a popular colleague but he also worked hard. At the Albany Museum he reorganized the collection of fossils, one of the best in the country, and at the College he was solely responsible for teaching geology at all three undergraduate levels. As a conscientious scientist Schwarz seldom passed a year without papers in the *South African Journal of Science* or the *Geographical Journal* and in due course he collected the accolades of a successful academic career: president of one of the sessions at the S_2A_3 conference in Grahamstown in 1908, fellow of both The (British) Geological Society and the Royal Geographical Society and eventually in 1923 president of the South African Geographical Society. But the long hours took their toll. As Schwarz became more desk-bound, his papers lost empirical content and became more speculative: the origin of the Witwatersrand banket (Schwarz 1911), the lost continent of Agulhas (Schwarz 1915) and faults in general rather than in particular (Schwarz 1916). He also discovered that text books eased his task of lecturing.

Two concise books, *Causal Geology* and *A South African Geology*, appeared from his pen in 1910 and 1912 respectively: the former discussing geological processes, structures and forms, the latter directed at their southern African manifestations.[2] Both were causal in approach but Schwarz went out of his way to present unorthodox explanations. "I have ...", he wrote in the preface of *A South African Geology* (1912:1), "put in some material which is still on the side of unrighteousness."

Three of these "injustices" had a direct bearing on his later visions. First, Schwarz accepted TC Chamberlin's so-called Planetismal Hypothesis which postulated that the Earth's sphere was formed by the aggregation of large meteoric lumps circulating around the sun. From that he erroneously concluded that the Earth's interior was cold and solid. Second, Schwarz asserted that the Earth's crust everywhere consisted of the same granitic material. Although the continents and ocean floors were litholitically similar, they were subject to a process which he called "secular uplifting and submergence". Sediments were continuously transferred from a continent to the bordering ocean floor and this disturbed the equilibrium between the two crustal fragments. Eventually, after a period of denudation, a particular continental fragment would sink beneath the sea, while, to restore the equilibrium, the adjacent oceanic fragment would rise to emerge as a new continent, a process of continuous oscillation. "Just as the continents have been covered repeatedly by the ocean, the ocean floor as often has been exposed as dry land" (Schwarz 1912: 23). For Schwarz southern Africa was the prime example of a recently uplifted ocean floor: The interior plateau was a succession of plains over which the incision of rivers had scarcely begun and its present cycle of denudation might still take another 9 Ma (Schwarz 1912). Thirdly, according to Schwarz the rains falling over the continents were only to a limited extent derived from oceanic moisture: their origin was the transpiration from plants covering the land. Therefore "the more cultivation there is and the more forest there exists on land, the more rainfall there will be ... We could in South Africa increase our rainfall by something like 70% if the wastelands were cultivated" (Schwarz 1912: 27).

Now, more than a hundred years later, Schwarz's ideas look bizarre but they reflected the geographical wisdom of the turn of the nineteenth century. Physicists claimed, on what appeared to be incontrovertible evidence, claimed that the Earth was only 20-30 Ma years old. Geomorphologists, therefore, underestimated the duration and complexity of land-forming processes. Little was known of the rocks and morphology of the deep ocean floors: the notion that the Indian Ocean was a sunken continental block was proposed by the respected Eduard Suess and also accepted by Passarge. Reliable techniques to measure the transfer of moisture from oceans to the continents were lacking and Earth scientists of the calibre of E Bruckner and A Supan supported the terrestrial origin of rains. Nevertheless, for most South African geologists *Causal Geology* went too far (Schwarz DM 1945).

The Union of South Africa's participation in the First World War brought tension among whites. English speakers supported the War enthusiastically and many of Schwarz's students joined up, some in British uniform going to Flanders, others among the South African who invaded German South West and East Africa. As student numbers dwindled, Rhodes University College experienced tight times but in the Eastern Province, resounding with calls for king, country and Empire, the vision of a Greater South Africa was vividly rekindled. At the same time Union Government was reforming tertiary education. Rhodes University College, if for other reasons than Stellenbosch, had similar doubts about sacrificing its identity to one national university and welcomed the compromise embodied in the three university acts implemented in 1918. In terms of Act No 12 of 1916 Rhodes and the other five constituent units of the federal University of South Africa acquired control over their own teaching and the freedom to initiate new subjects. Among the latter was geography, the 1915 college calendar announced the introduction of the subject the following year and just as at Bloemfontein and later Potchefstroom the subject was assigned to the professor of geology. Schwarz, in a way, was prepared for the new responsibilities. Over the years his interests widened to ethnography, astronomy and the history of discovery. Geology students at Rhodes tried to make the best of Schwarz. As a lecturer he was fatherly and fair and guided them safely through the peaks and pitfalls of the discipline, although some complained that he walked them off their feet during excursions. They found him serious and preoccupied in his own affairs, just the sort of person to be gently heckled. One student wrote (*The Rhodian* 1920: 17).

> Geography, geography?
> It's such a pleasant study.
> It tells you why the ocean is dry
> And why the desert is muddy.

In 1917 he started collecting information for a textbook in geography, eventually published in 1921, with a regional focus on Africa (EDM 1968). He skimmed through English and German geographical literature, glanced at the coarse maps of Africa and noted the fossilised drainage channels, dry lakes and pans of the Kalahari which Passarge interpreted as symptoms of recent desiccation. Remembering his walkabouts in the eastern Folded Belt and the many conversations with Kouga farmers, he accepted desiccation as a real and immediate threat. Digging in *Causal geology* for explanation, and scanning southern Africa in his mind, his ideas blended into a vision which not only explained the problem of drought but promised to solve it, a master plan to rehabilitate the pauperized farmers of the Karoo and construct lifelines for a Greater South Africa.

The Scheme: "Restore those lakes"

Schwarz introduced his vision in a brief article in the weekend edition of the Johannesburg evening paper *The Star* of 31 January 1918. Geological and historical evidence suggested, he claimed, that 300 years earlier southern Africa had a rainer climate. Rains during those times were fed by moisture which had evaporated from two lakes in the northern Kalahari. Since then the lakes had been transformed into the desolate and glimmering Makgadikgadi and Etosha Pans. One consequence was the scorching westerly winds which buffeted the Highveld and Karoo. The lost paradise might be partly regained by constructing weirs across the Chobe and Kunene Rivers, returning their flows to the original depressions and creating an open water surface with the extent of Lake Tanganyika. Unless this happened, southern Africa was on its way to become a second Sahara. Schwarz (1918) asked the public's support to avert this disaster.

Within weeks Schwarz's appeal echoed all over South Africa. Farmers and organized agriculture seized the vision and the press condensed it into one clear call: "Restore those lakes." Politicians had to listen. On 12 March 1918, not two months later, the Union Senate (Debates of the Senate 1918) accepted the following proposal:

> This house wishes to draw the attention of the Government to the scheme proposed by prof Schwartz (sic) ... claiming to result in the refilling of the Etosha and Makarikari Lakes and restoring the climatic conditions, as far as moisture is concerned, ... to the whole of South Africa.

Schwarz, then 45 years old, found a cause and went on to propagate his scheme at all levels. In an address presented to the 15th conference of the S_2A_3 in Johannesburg in July 1918 he framed his vision within a scientific context but also had to fend off the first probing questions (Schwartz 1919a). After a trip to the Etosha Pan and Ovamboland (Figure 1), which still displayed vivid scars of a drought three years before, follow-up papers appeared the next year (Schwarz 1919b, c); a substantial book, *The Kalahari, or Thirstland Redemption,* came off the press early in 1920 and, later during the year, Schwarz went to London to argue his case before the African Society (Schwarz 1920b) and the Royal Geographic Society (Schwarz 1921a). In the discussion which followed the latter presentation criticism was harsh and pointed. Clearly the scheme was controversial. Before covering the debate in Sections 4.4 and 4.5, it is worth looking at the details of Schwarz's arguments as they were developed in *The Kalahari*, his most comprehensive exposition.

Late eighteenth and early nineteenth century travellers, Schwarz asserted, described the southern African interior as a succession of waving grasslands with perennial rivers, wetlands and vast herds of game. By the 1820s Lake Ngami, the last of the open water surfaces, had dried up and droughts started ravaging the plains of the Kalahari and eastern plateau. While the immediate cause was the dry westerly winds, the fundamental cause was the crustal process of "secular uplifting and submergence".

During the Early Tertiary the crustal fragment which now formed the African continent was thrust upwards from the ocean floor, emerged from the sea and became a plateau, but with a succession of marginal faults along the coast, facing outwards as mountainous ramparts (Schwarz 1920). In the interior of the continent the rapid uplift caused shallow down warps which in the north, centre and south were soon filled by extensive lakes overflowing along rivers running leisurely for thousands of kilometres to distant outlets. A Proto-Niger, Proto-Congo and Proto-Nile drained northwards to the Gulf of Syrte in the Mediterranean Sea, while a mighty Proto-Orange flowed for 2 000 miles (3 200 km) south-westwards, first filling the Makgadikgadi Depression then continuing along the Old Molopo and Lower Orange to discharge into the South Atlantic Ocean. Initially, all the lakes provided moisture for copious rains over the continental interior but the rivers, and in due course also the lakes, were tapped directly towards the ocean by short, fast-flowing, headward-eroding streams which rose on the steep outer slopes of the ramparts.

Over East Africa the lakes were still intact and the Nile, their outlet, shifted only slightly eastwards from its original position. But further back in time the Proto-Niger and Proto-Congo were diverted directly to the Atlantic Ocean and their lower-course environment degenerated into the Sahara Desert. The dismemberment of the Proto-Orange took place in historical time. Schwarz quoted 16th century Portuguese missionaries (Schwarz 1920c) who placed the origin of the Zambezi just downstream from the steep head of the present Victoria Falls. In other words, before 1500 there was no falls and the upper course of the present Zambezi drained directly into the Makgadikgadi Depression. In the west the Kunene River was being caught in a present act of piracy. Under conditions of low and average run-off the river already tumbled across the Ruacana Falls to the South Atlantic Ocean but during high floods it spilled over the low watershed towards the east, flooding Ovamboland and ending in the Etosha Pan.

Schwarz's conception of Africa's geomorphological history presented him with a solution. While the master stream of the Zambezi was irredeemably lost to the Indian

Ocean, weirs of 10 and 45 feet (3 and 15 metres) in height in the Chobe and Kunene Rivers respectively would restore run-off to the central and downstream sections of what was originally the Proto-Orange system. The Chobe weir, just upstream of the present tributary and where its associated marshes approached the Zambezi, would impound water south-westwards towards the present town of Maun and then south-eastwards into the Makgadikgadi Depression. The Kunene weir would divert water from the present middle course of the river across Ovamboland southwards to the Etosha Pan, then eastwards along the Omuramba Omatako and Okavango River to Lake Ngami and from there to the Makgadikgadi Depression. A restored Greater Makgadikgadi Lake would transform the subcontinental interior:

> A sheet of water this size ... in the middle of the Kalahari would turn this thirstland from being a source from whence all drought producing hot winds originate, into an evaporating dish supplying clouds for South Africa, ... for moisture ... as it is transpired by the leaves of forest trees and the grass of meadows, or evaporated from the surface of the lakes, ... ascends into the air and is precipitated as rain over the adjoining country, and the process is repeated in ever widening circles (Schwarz 1920a: 38).

The availability of surface water and the ameliorated climate would favour large-scale irrigation. Schwarz estimated that 15 000 to 20 000 white families could be settled on 600-acre (240 ha) plots along the restored Omuramba Omatako and Lower Orange channels to cultivate tropical crops. Ovamboland might remain an African agricultural area, but in northern Bechuanaland reserves away from the irrigated areas should be allocated to the local people.

The Kalahari Scheme's benefits, Schwarz believed, were threefold: the poor-white problem was addressed comprehensively; construction would be cheap because storage and distribution exploited pre-existing natural features and; the surfaces available for evaporation and transpiration would comprise 100 000 square miles (161 000 km^2), one fifteenth of the surface area of the subcontinent. After ten years the increase of atmospheric moisture would augment the average annual rainfall of the subcontinental interior by 10 inches (250 mm) (Schwarz 1920a).

The debate on the Scheme soon acquired self-momentum. It was mooted again in the Union Senate in February 1919 and in the Assembly in 1920. Public opinion polarized. Pure and applied scientists were sceptical, the public enthusiastic.

Contesting the image: Scientists vs amateurs

Schwarz's Kalahari was an imagined place. While he was well acquainted with the Cape Fold Belt, he saw a Kalahari environment for the first time when he visited Ovamboland in October 1918, ten months after his article in *The Star*. Apparently, he never reached the Kunene River (Du Toit 1926). The Scheme was a blend of impressions, perceptions and speculations braided together in Schwarz's study and projected on a map at the resolution of a first-year textbook. The scientific framework was abstracted from *Causal Geology* and *A South African Geology* but by 1920 the books were clearly outdated. Wegener's theory of continental drift of 1912, though still controversial, provided a clear alternative to "secular submergence and uplifting", while the first radiometric dating of rocks, pioneered by Bertram Boltwood in 1907, suggested a geological timescale one hundred to a thousand times longer than physicists assumed. Moreover, Schwarz used published sources indiscriminately. Travellers' tales of abundance were emphasized, those of scarcity ignored (Potgieter 1929). The hypsographic data on contemporary maps hardly supported an impoundment of the Chobe River to reach back 400 km, while Schwarz was seemingly unaware of Passarge's Kalahari *Schwelle*, which divides the Makgadikgadi Depression from the Molopo basin. "Over and again the boldest conclusions from very inadequate data", was Serton's (1928: 360) dry comment.

Contemporary scientists also criticized Schwarz on details. His most powerful early opponent was FE Kanthack, then Secretary of Irrigation and civil engineer by training. At the July 1918 conference of the S_2A_3 he pointed out that the Molopo basin had a long history of aridity and questioned Schwarz's naïve assumptions about the engineering side of the Scheme. At the Royal Geographical Society in November 1920, Kanthack produced new data from the Kunene Boundary Survey of which he was a member (Schwarz 1921a). The Kunene upstream of the Ruacana showed no sign of overtopping its watershed, the hydrographic gradient across Ovamboland was too flat to allow water to spread southwards and, even if this should happen, the total run-off of the river was only sufficient to fill the Etosha Pan to a depth of one metre. These arguments sunk the western sector of the Scheme. The meteorologists, Stewart at the S_2A_3 in 1918 and Gold at the Royal Geographical Society two years later, were less adamant and conceded their incomplete knowledge of the climatic processes influencing the African interior but on one point they agreed: a higher moisture content over the plateau would not by itself ensure higher rainfall.

In the middle of the decade during which South African scientific enquiring reached maturity, Schwarz's colleagues found him awkward in debate. He did not so much refute their criticism as to repeat his assertions paper after paper, sometimes adding anecdotes from supporters which illustrated but did not explain. As the 1920s wore on many scientists saw no point in carrying the debate further. The onus was on Schwarz to produce new evidence and until that happened, they ignored him. In his presidential address to the S_2A_3 in 1923, AW Rogers (1923), Schwarz's erstwhile colleague at the Geological Commission, considered the parameters by which palaeoclimates could be reconstructed and concluded that South Africa's climate since the Cretaceous varied within narrow limits around the present. He clearly contradicted Schwarz but never mentioned him. Similarly, the Drought Investigation Commission of 1923 found "no proof of appreciable change in the Union's average annual rainfall in recent historical times" (UG 49-23: 5) but suggested that inappropriate farming methods turned successive years of below-average rainfall into a drought. The report did not refer to Schwarz which was apparently a sore point for him (Duerden 1923).

Schwarz was a visionary in the popular mind. According to the Cape Town morning newspaper *Die Burger* (8 April 1925), he "was someone of serious disposition...an exceptionable student who tends to do research...and does not give flippant opinions", but one who had empathy with the ordinary person. What *Die Burger* did not say, was that Schwarz stamped his scientific authority on long-standing myths which people wanted to believe. Since the 1820s there were farmers who claimed that rainfall was decreasing with the Eastern Province suffering serious droughts in 1903/04, 1912/13 and 1918/19. In 1914 the *South African Agricultural Journal* devoted one of its numbers to the theme "Is South Africa drying up?" and government investigations in 1914 and 1916 kept the problem in the public eye.

Schwarz projected these concerns onto a particular region. He saw the Kalahari as the geographical core from which the desert was expanding eastwards, a process that was not irreversible. There was the belief that the threat could be blocked by irrigation, dry-farming and tree-planting (Beinart 2003). South African rivers, some claimed, were running to waste in the oceans. The alternatives was to conserve them on land, turn them on the fields, allow crops and trees to enhance the water vapour of the atmosphere. William Macdonald (1914: 5), editor of *South African Agricultural Journal* captured these assumptions in a 1913 editorial:

> Is it possible ... to check the advance of the desert, conquer the crop blighting winds of aridity, and ameliorate the climate conditions of a vast country the size of the Kalahari? Yes, but three things are essential – Population, Conservation and Afforestation.

The greatest supporters of the Scheme were the amateur scientists, the self-made contractors, builders, miners and farmers who had a practical interest in the environment. Between 1918 and 1920 they rushed into the debate, clamouring in the correspondence columns of newspapers and periodicals on their fictions and superstitions of rain-making. Put the drylands under the plough, bombard the clouds with artillery, unload the voltage between the Earth's surface and clouds with kites! One correspondent proposed that the Greater Makgadikgadi Lake should be named after Schwarz (Christie, *Farmers' Weekly*, 25 December 1918), another offered to construct the weirs (Schramm, *Farmers' Weekly*, 19 February 1919) and not a few claimed that the Scheme actually originated with them (Deacon, *Cape Times* 26 February 1919).[3] The scepticism of Schwarz's fellow scientists did not upset the Scheme's supporters. At best the professionals were one-sided because "the commission under Mr Kanthack reported only on the engineer's side of the problem", at worst they were prejudiced, "people who already made up their minds when they opened the investigation" (*Die Burger*, 8 April 1925). Though many amateur scientists were too individualistic to swallow the Scheme hook, line and sinker (Weidner 1925), they admired the daring of the pioneer and lauded what they regarded as the boldness of Schwarz's vision. "In the midst of so much conservatism of thought", the *Farmers' Weekly* (8 November 1918) editorialized, "and so little disposition to think big, we can but recommend an attempt at so grand a scale to solve the greatest of our problems."

While popular enthusiasm was insufficient to launch a costly project, Schwarz had another and more subtle card to play: Greater South Africa. South West Africa had just become a Union mandate and a large irrigation scheme in Ngamiland might be the knot to tie South West Africa, Bechuanaland and the Rhodesias into a northward-extending Greater South Africa. Apart from poor whites, demobilized soldiers could also be settled on the Scheme (Schwarz 1920b). During his visit to London in late 1920 Schwarz used his contacts in the African Society to solicit support for a Kalahari Development Company to use "the northern rivers and lakes". In a letter to Leo Amery, Undersecretary for Colonies in Lloyd George's Coalition Government, he asked for an investigation of the Scheme and, should it be found feasible, a grant of land to start development. Amery referred the enquiry via the British High Commission in Pretoria to the Union and Bechuanaland Governments.

The response was negative. In his reply of 21 January 1921, Prime Minister Smuts, already used to deflecting questions on the Scheme, expressed the doubts of his Government's professionals:

> So far as the Union's territory and Bechuanaland are concerned, in the increase in rainfall which Prof Schwarz promises will result over the whole of South Africa, he has certainly not produced sufficient argument to show that the result he anticipates will follow if the Scheme be carried out.

Bechuanaland officials were more outspoken. RW Stigand, resident commissioner at Maun and the foremost expert on Ngamiland, dismissed Schwarz's sources on the area as "antiquated, useless and misleading" and questioned the competence of the "delta" spillways to transmit run-off to the Makgadikgadi Depression. The only reliable map was Passarge's, based on measurements taken 20 years before. Canalizing the spillways would drain and dry out the soils on which the Batswana depended for their cultivation of millet.

The Union Government acted circumspectly. Knowing its voters' desire for irrigation projects, the Scheme was referred to the Advisory Board of Trade and Industry in 1921 where it was wrapped away pending funds becoming available for a feasibility study. The political attractiveness faded when Rhodesian voters declined to join the Union in 1922, a decision which Schwarz deplored and blamed on the Union's political leadership. Look north, he told geographers in his presidential address to the South African Geographical Society in 1923. "Rhodes had the geographical sense developed instinctively ... and dreamed of a vast railway traversing the country from north to south, so that South Africa ... should be able to reach out into the interior and by her commerce derive those benefits to which her sacrifices in the cause of civilization entitles her" (Schwarz 1923a: 6).

The image fades

In June 1924 a coalition of the National and Labour Parties led by Hertzog ousted Smuts' South African Party Government and Little South Africans came to power with politicians who were unenthusiastic about a Greater South Africa but sensitive to the farmers' lobby. On 7 April 1925 FJ Du Toit, the nationalist MP for Victoria West, tabled a motion in the Assembly requesting the Government to investigate the feasibility of the Kalahari Scheme. In the ensuing debate members of all parties representing Karoo constituencies supported Du Toit and the request was carried unopposed. After a private donor contributed an additional £1 000 towards the cost, the Prime Minister authorized a reconnaissance and Schwarz was invited to accompany the researchers in his private capacity.[4]

The expedition was organized by the Department of Irrigation, with the geologist AL Du Toit, Schwarz's successor at the Geological Commission, as leader. With Livingstone in Northern Rhodesia as point of departure, the team traversed northern Bechuanaland and the adjacent Caprivi Zipfel in South West Africa and, using theodolites and aerial photogrammetry, mapped the Zambezi and Chobe Rivers upstream from Victoria Falls. Du Toit's report of May 1926 (Department of Irrigation 1926) mercilessly deconstructed Schwarz's Scheme. First, Du Toit found no evidence of a higher rainfall or a Greater Makgadikgadi Lake in historical times. Run-off from the Okavango River did reach the Makgadikgadi Depression in the *geological* past, but flow became blocked by the build-up of the "delta" upstream of a south-westward striking subsidence more or less coinciding with the line of the Thamalakane channel. The dry lake was, therefore, not caused by stream piracy at Victoria Falls but by tectonic movement. Second, the Chobe's run-off was so limited that irrigation could only be considered with augmentation from the Zambezi and even then water would be sufficient for only 600 000 acres (243 000 ha), a full hundred times less than Schwarz's estimate. Finally, Du Toit rejected the notion that evaporation from an open water surface could enhance rainfall over the Kalahari and adjacent parts of the Union. Already half the run-off that Schwarz wanted to turn into the Makgadikgadi Depression was evaporating in the northern Kalahari with no effect on rainfall.

For Schwarz the expedition was another opportunity to gain first-hand knowledge of the Kalahari, now in association with colleagues, but he preferred to go his own way. Both Du Toit and Schwarz started from Livingstone but Schwarz wanted to see Lake Ngami and the Makgadikgadi Depression while Du Toit focused on the nearby Zambezi. Ironically, for someone who made desiccation the central prop of his Kalahari image, Schwarz found the spillways to the south in full flood, had to complete part of his journey by boat and suffered the indignity of having his dugout canoe dumped on a flooded field of maize (Schwarz 1928). He was unperturbed. "If my project has been in working order", he wrote to the editor of *The Star* (Schwarz, 11 August 1925), "the whole of the flood water, instead of going over the Victoria Falls to the sea, would have gone into Makarikari and a lake as big as Tanganyika would have been formed."

The Du Toit report, which was scathingly critical of Schwartz as a scientist, came as a severe shock especially since the two geologists took leave of each other in June 1925 in great friendliness (Letter Du Toit to Schwarz, 13 September 1925). Withdrawing under criticism, Schwarz avoided Du Toit and chose the theme for a most readable book on his Kalahari journey (Schwarz 1928) not palaeoclimatology or hydrology but the area's inhabitants. He could not disprove the hard facts of the expedition's conclusions but was determined to uphold the integrity of his geological concepts. Some years earlier he mounted the University College's 9½-inch telescope in his garden in Grahamstown, meticulously studied the moon's surface and noted features which he regarded as similar to the "fault-pits" of the Baviaanskloof. "Imbricate faults", he thought, were small-scale manifestations of the massive process of "secular submergence and uplifting", his presumed key to southern Africa's geological history (Schwarz 1923, 1928).

In 1927 Rhodes University College appointed a second staff member to share the teaching of geology and geography with Schwarz and the following year the latter took sabbatical leave to search in West Africa for confirmation of his theories of continental oscillation and stream piracy. Six weeks before he was due to return he contracted enteritis and died unexpectedly on 19 December 1928 of heart failure in St Louis (Senegal) at the age of 55. With great sacrifice Daisy Schwarz went to recover her husband's remains and had him buried in London. In her eyes he was a martyr, "one who gave his life to show us a vision of a greater South Africa" (Schwarz DM 1928: 620). Over the next twenty years she kept Schwarz's memory alive in popular magazines until he became greater than life itself: *the* Plan, *the* Scheme, *the* Professor, *Schwarz of the Kalahari.*

Each time South Africa was hit by drought during the next twenty years, the Scheme was resurrected. In the middle of the Great Depression and drought in the early 1930s a group of Free Staters assembled a Kalahari Redemption Society and in 1935 Deneys Reitz, Union Minister of Irrigation, paid a hurried visit to northern Bechuanaland, once again to dismiss the Scheme (Reitz 1999). A further drought in 1943/44 and the possibility of settling South African ex-servicemen in the north-east of South West Africa after the Second World War, drew the Union Minister of Lands, AE Conroy, to the area in 1945. The consequent report (UG 28/46) confirmed Du Toit's conclusions and supplemented them with an insightful review of the climate of the southern African interior. Politically, however, the Minister bungled by veiling the visit in secrecy and inviting nine MPs of his own United Party along which was in the eyes of the Opposition, a hunting trip to a remote and auspiciously game-rich region (*Die Burger* 25 June 1945). The Scheme flickered for the last time and finally disappeared from the screen.

A dead end and new beginnings

In the literature of South African geography, the works of Passarge and Schwarz stand in subtle contrast. Both made errors of interpretation and both were widely criticized. Passarge's Kalahari was real and had an empirical base which could be pursued and improved within the framework of the Earth Sciences. Conversely, Schwarz's trajectory advanced him from an empirical Cape Fold Belt to an imagined Kalahari. His personal knowledge of the area was based on two fleeting visits and his argument was developed from outdated concepts to further a cause for which he appealed to the public rather than to his scientific peers. The Scheme was an intellectual dead end.

Yet, Schwarz stimulated new departures. He drew attention to the northern Kalahari, a relatively inaccessible area which otherwise would have remained terra incognita for decades. Still, in the 1920s expeditions by Stigand (1923) and Clifford (1929) went beyond Passarge and Schwarz with new empirical information and shortly afterwards the multidisciplinary Vernay-Lang expedition (Rey 1932) revealed the exceptionally rich animal life of the Okavango "delta", a far richer and more sustainable resource than irrigation. In South Africa Schwarz controversy deepened the environmental discourse, South Africans became more aware of the uniqueness of their environment and they took the first serious steps to conserve it. The Kruger National Park, after all, was proclaimed in 1926. The controversy also permeated the seminal geographical discourse and drew responses from Serton (1928), Plummer (1926, 1929) and Wellington (1926, 1955).

In Grahamstown the Geology Department and its Geographical ward expanded and in 1936, eight years after Schwarz's death, Geography went its own way and through the work of JVC Rennie (1902-1989), trained as a geologist and a later Vice-Chancellor of Rhodes University, grew into a centre of excellence on its own (Fox & Daniel 1989). Schwarz's publications on the Cape Folded Belt are still valuable pathfinder works and the high-level surfaces he identified were integrated into the comprehensive models of later geomorphologists. And in the 1980s, Earth scientists found evidence of an even Greater Makgadikgadi than Schwarz had surmised, but 2 Ma earlier in geological time.

Endnotes

1 Chapter 4 has a predecessor in a paper by the author published in the *South African Geographer* (Barnard 1992). While the 1992 paper focused on the Scheme as such, Chapter 4 is concerned with the Scheme as a construct formulated by a particular individual. Archival sources on Schwarz are, however, limited. Previously unpublished information was extracted from the Schwarz file at the Cory Library of Historical Research at Rhodes University, Grahamstown and correspondence on the Scheme between 1920 and 1925 in the files of the Governor General of the Union of South Africa in the Pretoria Archives Depository. Perusasl of reports, reviews, correspondence and editorials in South African newspapers and periodicals between 1918 and 1919 provided useful insights into the controversy surrounding the Scheme.

2 Schwarz regarded *Causal Geology* (1910) as his magnum opus. According to (a probably apocryphal) rumour, he suspensefully awaited the first consignment from his Edinburgh publisher but great was his dismay when the books were unpacked in Grahamstown and he read the title, printed in gold on a dark blue background, misspelt as *Casual* (sic) *Geology*!

3 To what extent the Scheme was authentically Schwarz or lifted from someone else is difficult to establish from the sparse references in his publications. Similar ideas had been mooted among settlers in German South West Africa ten years earlier regarding the Kunene River and the Etosha Pan, notably by F Gessert and HJC Paulsmeier (Smit 1991). Schwarz quoted Gessert at length in *The Kalahari*, Gessert clearly being an ardent supporter of the Scheme. Even So, Schwarz was adamant that the overlap of his ideas with those of the Germans was coincidental (*Farmers' Weekly* 18 February 1919). Certainly, his vision was more comprehensive – he also brought the Chobe into the Scheme – and argued within a supposed scientific framework.

4 In 1925 Schwarz apparently found an intercessor with the Pact Government in the person of EH Louw, the recently elected National Party MP for Beaufort West. Louw studied law at Rhodes University College between 1912 and 1918 and, in the small compact university community, was probably well acquainted with Schwarz. Louw was South African Minister of Foreign Affairs between 1955 and 1963.

CHAPTER

5

Serton, human Geography and Apartheid

Passarge opened the South African discourse of geography and Schwarz took it into the public arena. Petrus Serton (1888-1963), while occupying the first chair of geography instituted at a South African university, gave it a professional beginning and continuity. Among the founders he was the odd man out: a human rather than a physical geographer; continental and not Anglo-Saxon in education, publishing in Dutch and Afrikaans rather than in English; covering the whole gamut of geographical genres from professional monographs, papers in journals to edited diaries, school textbooks and newspaper reviews.[1]

Serton was the only founder who worked and lived in an Afrikaner milieu. At first sight his career trajectory looks starkly simple. Born and educated in the historic city of Utrecht in the Netherlands, he worked for only three years in Rotterdam before accepting an offer from the University of Stellenbosch where he settled in 1920 and lived until his death in 1963. While Stellenbosch remained as a town over these 43 years, the mood of the university community followed the ideological stance of Afrikaner nationalism at large: searching during the 1910s and 1920s, discordant and rancorous in the 1940s and arrogantly self-confident during the 1950s and 1960s. Serton's output as a geographer reflected these shifts: supportive during the 1920s, sidelined and embittered during the 1940s and critical of apartheid in the 1950s. Before taking up the Serton narrative the appointment of Serton begs one question: Why did Stellenbosch, in contrast to other university colleges, establish a chair in human and not in physical geography?

Stellenbosch opts for human geography

Stellenbosch, 60 km from Cape Town and South Africa's second oldest town, became an educational centre in 1859 when the influential Dutch Reformed Church, suspicious of the liberal ideas with which young ministers returned from the Netherlands, established a theological seminary in the graceful old Drostdy building on the northern bank of the Eerste River. In 1866 the Stellenbosch Gymnasium, the local boys' school, started preparing senior pupils for degree examinations and in 1866 the Stellenbosch College hived off to specialize in undergraduate teaching, rechristening itself the Victoria College in 1887. With the Theological Seminary next door, the campus sentiment was Cape Dutch and conservative. Yet, for most of the nineteenth century Cape Afrikaners lived contentedly under Queen Victoria's rule, the few professors at the College were sober Scottish Presbyterians and their language of instruction English. This spirit of tolerance was rudely shattered by the Jameson Raid and the South African War which made the defeated Transvalers, Free Staters and their Cape Dutch sympathizers aware of a common Afrikaner identity.

At Union in 1910 Victoria College, the largest of the university colleges under the University of the Cape of Good Hope, was a bastion of Afrikaner nationalism still uncertain about its goals. While Louis Botha's governing South African Party stood for a merger of Afrikaner and English whites, Hertzog's National Party of 1914 distrusted Botha's closeness to Empire and his indulgence towards the English language. Unsurprisingly, Victoria College perceived the proposed national teaching university on the English model as a threat, a nursery for capitalists, imperialists and "imitated Oxford fops with impeccable English accents" (Brummer 1918: 198). As the university controversy spread countrywide between 1912 and 1916, Victoria College fought back vigorously, organized the support of old students and stressed its Afrikaner identity. "Stellenbosch stands for an ideal", argued a memorandum presented by three alumni to the Minister of Education in 1913 (Malherbe 1918: 68). "For the Afrikaner people (*volk*) it represents more than just an educational institution. It is a symbol of and guarantee for a vigorously growing national life searching for self-expression." This search, however, was diffuse: the University Chancellor and Rector were in favour of maintaining Dutch, the stilted language of culture, while the "Young Turks" on campus wanted to develop Afrikaans, the crude but flexible colloquial. There was also a sound academic reason for opposing the concept of one national university: a central institution of higher status might siphon off the best students from the lesser colleges for its postgraduate courses. Victoria College's strong opposition and the eventual compromise embodied in the three University Acts of 1916, also enabled English-language institutions such as the South African School of Mines and Technology, and Rhodes University College, to maintain their identity.

While the Victoria College Council negotiated with the Department of National Education, the Senate attended to more mundane matters such as the training of teachers, among whom voices were being raised for a "new geography" at school. In 1913 the College had a teaching staff of 20 and 367 students. The Scottish professors were still there but they were being joined by Netherlanders and young Afrikaners with Continentally acquired PhDs. The Scots, mostly natural scientists, took the initiative to find a niche for geography. Early in 1913 they persuaded the College Senate to offer a one-year course in geography and later during the year got the University of the Cape of Good Hope to recognise geography as a BA Pass subject. Prof SJ Shand, an igneous petrologist of international standing and the driving force behind the proponents, emphasized the educational value of geography and compiled a syllabus heavily slanted towards physical geography. Implemented in 1914, the course was structured in three blocks. Shand was responsible for physical geology, climatology and cartography during the first two quarters; Prof FJ Goddard, a zoologist, used the third quarter for biogeography and ethnography and Prof W Blommaert, a Flemish and historian, lectured on political geography and the history of discovery in the final quarter. Moreover, the College showed its support by appointing a lecturer, Marguerite le Roux, to teach both geography and history.

With geography also expected to become a secondary school subject, Shand's course was popular although its block structure and the divergent backgrounds of the lecturers proved uncomfortable. For three consecutive years – 1916, 1917 and 1918 – the College Council unsuccessfully petitioned the Union Department of Education for funds for a chair of geography. In the meantime the options in geography were extended to two levels: a one-year course for second-year education students and a final-year course for degree students, both with Geology 1 as prerequisite. In April 1918 the University Senate went a step further by stipulating that the geographer they intended to appoint would be an all-rounder but with specialization on the sciences side. However, two unexpected crises occurred later during the year. First, Shand, still a British citizen, was called up for military service during the last months of the First World War, then the remaining lecturers, miffed by the University's refusal to pay extra remuneration, confronted the Council with an ultimatum: unless a full-time professor of geography was appointed, they would present the course for only one more year.

By 1919 the prospective chair of geography became entangled with the University's ambition to establish a Faculty of Commerce. For the Stellenbosch academic community the newly acquired university status was an opportunity to extend its mission: defining an Afrikaner identity, maintaining the "Hollands" languages and wiping out Afrikaner poverty. During the War years the first Afrikaans firms were established – Nationale Pers in 1915 and Sanlam in 1918 – and a director of companies, JG van den Horst, joined the University Council. Outside, however, the world of finance, commerce and industry remained an English preserve. "It is generally accepted", lamented the *Stellenbosch University Magazine* (*Stellenbosch Universiteitsblad* 1920: 6), "that a tough battle still awaits the Afrikaner on the economic front... [It] is as if nobody really knows how the struggle should be organized. One reason for this is undoubtedly the fact that attention at school and university is not attuned to the economic problems which pupils will be facing later in life."[2] Stellenbosch, the self-acclaimed Afrikaner "people's university" (*volksuniversiteit*), took the initiative and obtained the go-ahead to institute a chair of Economics. The Union Department of Education, however, blocked the University's candidate, JFW Grosskopf, because he was a 1914 rebel and as such still barred from a teaching position.

While still anxious to start a Faculty of Commerce and uncertain about Grosskopf, the University discovered a new niche for geography. On 7 November 1919 the Senate decided to institute a chair of geography, if necessary financed from its own funds, and to accommodate the discipline in all three undergraduate years. The person to be appointed should be someone "who has undertaken a complete course in the subject but with specialisation in commercial geography" (Senate Minutes 1919). The 1919 decision was a complete reversal from the previous year but there was no one to argue the case for physical geography. Shand, geography's mainstay since 1913, was just back from military service and still had to find his feet and Goddard was planning an Antarctic expedition which never came off.

Figure 3: Piet Serton (Oil on canvas portrait painting by Willem Pretorius, 2016)

Obviously, a commercial geographer was not to be found in South Africa and the University had a good reason to start its quest in the Netherlands. The Jan Marais Fund promised support for the chair on condition that half the courses should be presented in either Dutch or Afrikaans. The Dutch consulate in Cape Town passed the Stellenbosch enquiry to Prof Jan Niermeyer, geographer at the Utrecht State University (*Rijksuniversiteit Utrecht*), who happened to have the person Stellenbosch was looking for: his student Petrus Serton (Figure 3). Serton had just completed a doctoratal dissertation on an economic geographical theme at Utrecht. He also followed courses in physical geography, and as a staff member of the Rotterdam Advanced School of Commerce (*Hogere Handelschool Rotterdam*, today Erasmus University), was well acquainted with the structuring of tertiary-level courses in commerce. The Stellenbosch University Council appointed Serton to the chair of geography in February 1920 and was relieved to hear shortly afterwards that its other anxieties regarding the post were solved. In March 1920 the Department of Education notified the University that it would fund a chair of geography and concurred in the appointment of Grosskopf as professor of Economics.

The economic geographer from Utrecht

Petrus Serton – first name Piet – was born on 17 April 1888 on St Catharine's Crescent just outside the central core of the city of Utrecht where his father was a grocer. With the Calvinist respect for education and a tradition of tolerance, late nineteenth century Netherlands was one of the few places in the world where social mobility was possible for a clever lower-middle class lad. In 1863 the liberal Dutch premier Jan Rudolf Thorbecke instituted a system of secondary citizens' schools (*hogere burgerscholen*) which turned teaching away from the traditional classics to modern languages and science, the knowledge base of modern industrial society. By the turn of the century these schools were fully operational and highly successful: between 1900 and 1914 the small Netherlands produced five Nobel Prize winners. Young Piet Serton shared the benefits of an open education and followed the long road from the Primary Citizens' School of the Reformed Congregation to the State Secondary Citizens' School in 1908. For a university degree, however, Greek and Latin were still prerequisites and Serton put another two years aside for the State Examination (*Staatsexamen*) before registering at the Utrecht State University in September 1910 at the age of 22.

Situated at the outer edge of the High Netherlands just where the most northerly of the present distributaries of the Rhine splits into the Old Rhine and the Vecht, Utrecht was since the eighth century, the see of a bishop who held secular sway over the marshlands of Holland on behalf of the Holy Roman Emperor. The university, a Protestant institution established in 1614, was together with Leyden and Groningen elevated to a state university (*rijksuniversiteit*) in 1876 and could therefore claim government funds. For Serton the university, scattered over the city core but with the administration building within walking distance from his home, must have been a familiar institution. What was new, was the presence of geography in the curriculum, an option available only since 1908.

With the Amsterdam Municipal University (*Gemeentlijke Universiteit van Amsterdam*) already having an occupied chair of geography since 1876, the discipline was just as old in the Netherlands as in Germany, but there was no pressure from above to diffuse it to other universities. In 1906 the Dutch government, wanting to improve geography education at school, gave Utrecht the green light to include the discipline in its curriculum, the second after Amsterdam. The University Senate, after some deliberation, decided to accommodate both geography's primary subdisciplines and appointed two professors: the German Karl Oestreich (1873-1947) for physical geography and Dutch Jan Niermeyer (1866-1923) for human geography (Heslinga 1983). Niermeyer, also editor of the *Journal of the Royal Dutch Geographical Society* (*Tijdschrift der Koninklijk*

Nederlandsche Aardrijkskundige Genootschap), came from an educational background and commanded respect both as teacher and author. In his human relations, however, he lacked discretion. "In every case that Niermeyer dedicated himself", a later commentator (Heslinga 1983: 328) remarked, "he made enemies as easily as friends". The narrowly focused Oestreich and the more holistic Niermeyer were soon at odds. Physical and human geography went their separate ways ending with divergent curricula in different departments and faculties.

Although Serton opted for human geography within an Arts curriculum, he followed and enjoyed the introductory courses in physical geography and distanced himself from the discord between the two professors. Later, after attaining the D Litt degree, he expressed his appreciation to Oestreich. "I have often experienced", he wrote (1919: ii), "how indispensable a thorough knowledge of physical geography may also be for a person who concerns himself with human society." Yet, it was Niermeyer's conception of geography which struck deeper roots. In his courses he tried to balance German *Anthropogeographie* against French *geographie humaine* but his sympathy eventually swung towards the French school with their emphasis on human-milieu relations. As happened to the French geographers themselves, this approach led to area studies – regional geography in the terminology of the time – and among Niermeyer's publications the Dutch East Indies (Indonesia today) was the most constant theme. His lack of research experience, however, blinded him to the pitfalls of this approach. Area studies based on "a geography of relationships" were not directed at problems and questions, the research methodology was not standardised and the texts were prone to degenerate into pat descriptions (Dietvorst et al. 1983: 160-174).

Meanwhile, the young Serton had to build a career. In 1913 he passed the candidate's examination in Dutch Letters with distinction and sat for the doctoral examination in 1916 which enabled him to register for a DLitt under Niermeyer. With the First World War in full swing, neutral Netherlands provided refuge for one million Belgians and was under dire political and economic stress. Serton was fortunate to obtain an appointment as Niermeyer's assistant at the Rotterdam Advanced School of Commerce where the latter lectured as an extra-ordinary professor of Economic Geography. This subdiscipline appealed to Serton and he started collecting material for a thesis on the port of Rotterdam. Among the new colleagues he became acquainted with Hendrik Blink (1852-1932), editor of the *Journal for Economic Geography* (*Tijdschrift voor Economische Geographie*, after 1948 *Journal for Economic and Social Geography*) with whom he maintained contact for the rest of his life. But all was not work. In 1916 Serton married Eduarda van Roekel – first name Warda – after an engagement of seven years. She had grown up in the house of Serton's aunt in Vlissingen in the province of Zeeland.[3]

For his thesis Serton obtained key data from unstructured interviews with leaders of industry, port managers, senior supervisors and union officials, he kept meticulous record of all his conversations and assessed their judgment against the background of stale port statistics. His conclusions were summarized in a monograph of 187 pages, *Rotterdam as Port for Mass Commodities* (*Rotterdam als Haven voor Massale Goederen*, Serton 1919), which he defended successfully at the Utrecht State University on 19 December 1919, the first doctorate in human geography to be awarded by a Dutch university and Niermeyer's only successful doctoral student. (Niermeyer died four years later) The task of the geographer, Serton (1919) asserted in the introduction, was to create an accurate image of the Earth's surface. This image was not visual but conceptual, a construct built by selecting the researcher's most important and typical facts and arranging them by their relationship with each other. In practice, the Rotterdam monograph commenced with the commodities handled in the port – grain, timber, coal, metallic ores – and concluded with dock-side activities. Reviews appearing the following year lauded the monograph for being concrete, down to Earth and disciplined, but found it unnecessarily fragmented. The reviewers congratulated Serton on his appointment at Stellenbosch and ventured the hope that he might bring honour to Dutch science (Blink 1920; Van Blom 1920). They got more than they bargained for.

At first sight Serton's decision to exchange the Netherlands for South Africa seemed an abrupt one. The Stellenbosch University Council made the appointment in February, the Senate was notified in early March, the offer reached Serton some weeks later and he must have accepted it in April 1920. He arrived in Table Bay on the *Armadale Castle* on 12 July 1920, just in time to start lecturing for the second semester on 28 July 1920.

Why did Serton come to Stellenbosch? Circumstantial evidence suggests both push and pull factors. In 1920 the Netherlands was still suffering from the after-effects of war, the port of Rotterdam with its stagnant German hinterland worst of all. Nobody knew when, or even if, prewar prosperity would return and how it would affect an institution such as the Rotterdam Advanced School of Commerce (Serton 1919). Moreover, few academic positions were available in a country where only two universities had departments of geography. South Africa, by contrast, was for many Dutch a romantic frontier. During the South African War their support for the Boers was unrestrained. "Their fight (against Britain)...", a later historian (Kossmann 1978: 425) wrote, "was proof of a revival of the Dutch spirit. The Boers were reconquering South Africa for the Dutch race". The Dutch felt that Afrikaners needed them and many responded. While Serton was not beyond such sentiments, he also had an important personal consideration. Both Warda Serton's parents had died of tuberculosis and struggled with the disease for three years during the war. The couple, still childless, judged that Warda might benefit from a three- or four-year stay in a dry climate. Their decision was vindicated: she recovered completely and they stayed for life.

Establishing a geography department: Courses and networks

Stellenbosch in which the Sertons settled in July 1920 was a small town of 7 000 inhabitants, within easy travelling distance of Cape Town but culturally a world apart. Although visitors dismissed the town as a sleepy hollow, it had charm. Situated in the basin of the Eersterivier and surrounded on two sides by towering mountains, its white-washed Cape-Dutch and Cape-Georgian houses were shrouded in the deep shade of oaks branching across the narrow red, gravelled streets, dusty in summer and muddy in winter. The town functioned as a central place for the wine and fruit farmers of its productive countryside, but its real economic base was education. Over the years the town attracted three high schools, a primary teachers' training college, a conservatory of music, and, of course, fed the Theological Seminary which started it all. While whites formed half the town's population, Africans were nearly absent. The other half of the population were "coloureds": the Afrikaans-speaking, largely Christianized labourers, tradesmen and small entrepreneurs of mixed Malay, Khoi and white descent assimilated into the local economy. Their scattered residential quarters were integrated into the town layout and, although they lived segregated lives, their qualified vote in municipal and national elections sometimes held the balance between contending parties, as happened in Stellenbosch in 1924.

The Sertons accepted Stellenbosch as it was and felt at home. Piet, 32, short, with straight dark hair and formal in manner, arranged for a family residence to be built at 46 Jonkershoek Avenue, on the edge of the new Mostertsdrif suburb. With the Lanzerac homestead in the middle distance and the Jonkershoek Mountains a bold backdrop, the view to the south was stunning. The Sertons never moved anywhere else. Here their four children – three daughters and a son – were born and here Piet died 43 years later.

As good Calvinists the Sertons joined the local Dutch Reformed congregation which honoured its European roots. The older ministers still preached their sermons in Dutch and the Dutch Bible and hymnal book still held sway. The university's affairs were conducted less formally than Serton was used to in the Netherlands (Serton 1926). The six Dutch and one Fleming on the staff of 55 lectured in Dutch. The dwindling number of Scots taught in English while most of the Afrikaners were switching to Afrikaans. The 650 students came from all over South Africa and they were one hundred per cent Afrikaners. They lived simply in university residences, loved sport and were spontaneous in their relationship with their lecturers. Serton liked them to the end of his career (Nel 1989). But they took politics more seriously than their Dutch counterparts and were naïvely idealistic about Afrikaner issues of the day; namely an authentic, if rather undefined, South African political identity; the eradication of Afrikaner poverty; and the development of an Afrikaans language and culture which could stand up against English. In this age of innocence Afrikaners regarded Netherlanders as their natural allies. "Also to [Professor Serton]", wrote the *Stellenbosch Student* of September 1920 (21: 72), "our hearty welcome. Although South Africa is still foreign to him, we hope that he would soon feel at home. We may not be countrymen yet but we come from the same stock."

Of the three tasks which Serton took on, the geography curriculum had first priority. With the discipline now within a department in the Faculty of Arts and Humanities, Serton shifted the focus of teaching from physical geography to area studies for which he took responsibility. The connection with the Faculty of Natural Sciences was abrogated and the Geology 1 prerequisite dropped. An introductory course in physical geography was assigned to Ms Le Roux who presented it, together with her courses in history, to first years until her retirement in 1932. In 1925 Serton's systematic specializations in economic and political geography crystallized at second- and third-year undergraduate level respectively. A master's course was directed at the task and method of the discipline and the "production" of a thesis. The content of the courses breathed the spirit of Niermeyer and was free of the constrictive environmental determinism which clouded contemporary South African courses (Wesso 1992). Area studies synchronized with the regional geography popular in Cape schools (Clarke 1949) and prospective teachers found Serton's courses a useful career background.

The second task was the establishment of Stellenbosch's Faculty of Commerce. On arrival Serton found immediate rapport with JFW Grosskopf who had taken up his appointment as professor of Economics just three months earlier and Serton probably had a hand in the imaginative proposals for a Faculty which served before the University Council on 14 August 1920. Three years later, when the first commerce students registered, geography was an option up to the final year for the BA Commerce degree and to the second year for the BComm students. Because Grosskopf, also an Afrikaans playwright of note, apparently cared little for administration, Serton guided the fledgling faculty for its first years between 1926 and 1930. As the more specialized commerce subjects struck root, geography's input shrunk to a first-year course in Economic Geography handled in 1932 by DJ Conradie, appointed as the second full-time member of the Department in 1929.

Serton's third task was a matter of conviction. Like many Dutch he sympathized with Afrikaners and in Stellenbosch he was introduced to South African political discourse by colleagues such as Grosskopf and SFN Gie, both sophisticated but ardent Afrikaner nationalists. Serton knew where he wanted to stand but had to make a sacrifice. As Afrikaner nationalism became more confident during the 1920s, the clamour for the Afrikaans *and* Dutch languages shifted to Afrikaans only. In 1925 the Union Parliament interpreted Article 137 of the South Africa Act in such a way that "Hollands" was equated with both Afrikaans and Dutch. In practice, Dutch rapidly fell away but Afrikaans as a medium of culture still lagged far behind English. The sober-minded Stellenbosch Dutch saw the apotheosis of Afrikaans as a natural process. "For we 'Hollanders' in South Africa", Serton commented in a letter to *Die Burger* (26 February 1926), "there is only one correct course to follow. We should co-operate fully in the establishment of an Afrikaner culture which we also love and of which we also may be proud." He put his words into action. In 1925 Serton was a member of the Van der Horst Commission which recommended the extension of the Port Elizabeth harbour and their report was the first government report (UG 13/26) originally compiled in Afrikaans. In 1928 he published the first Afrikaans paper in the *South African Geographical Journal* and during the 1930s assisted with the writing of Afrikaans school textbooks for the Cape Province. He also considered it important to acquaint Afrikaners and Dutch persons with each other. Serton's popular reviews in *Die Burger* and *Die Huisgenoot,* always in Afrikaans, were often comments on the Dutch scene, while the more formal papers in Dutch in the *Journal of Economic Geography* (the present *Tijdschrift voor Economische en Sociale Geografie*) invariably focused on South Africa.

For the professional geographer specializing in area studies, South Africa was a fallow field and Serton grasped the opportunity. He liked themes which reflected the subcontinent's individuality and his papers were sometimes fine and at other times coarse in resolution, namely on the one hand the port of Cape Town (1923) and farming in the Laingsburg district of the harsh western Great Karoo (1929) and South Africa's diverse population (1923, 1925) and its railway system (1929) on the other. The two perspectives were also conflated in concepts such as the meaning of distance (1928a) and desert (1926, 1928b 1929).

In these studies Serton remained faithful to his original guidelines: the geographer's task was to create an accurate image of the Earth's surface. First, he had to acquire a thorough personal knowledge of the area he wanted to study and in this respect Serton displayed an utter disregard for comfort. In 1925 he travelled the High Drakensberg by pony, in 1926 he journeyed to remote parts of the Laingsburg district by accompanying the local doctor and later, in 1935, crossed the Kalahari, for at least part of the distance, on the back of an open truck. Second, because detailed maps and statistics were not yet available, he collected his information from well-informed locals: in Laingsburg for example, from the magistrate, the mayor, the town clerk, an attorney, the school principal, experienced and successful farmers, stock inspectors and so forth. Last, this information had to be ordered and interpreted. For Serton, people were the measure. "The economic development of a country," (Serton 1929: 67) wrote in a book review in the *Journal for Economic Geography,* "is not simply a consequence of its natural circumstances; human factors are at least as important. The richest natural resources are valueless unless people have the knowledge to put them to use." This was not the language of an environmental determinist.[4]

By the end of the decade Serton's work received recognition outside Stellenbosch. He served the 1928 term as president of the South African Geographical Society, presented a paper during the joint conference of the S_2A_3 and the British Association in 1929 and travelled the United States on a Carnegie grant in 1931, an opportunity to meet some of the leaders of the first generation of American geographers, among them William

Morris Davis (then at Columbia and still remembering his visit to South Africa 26 years before), Armin Lobeck also at Columbia, John Leighly at Berkeley, Harlan Barrows at Chicago and Derwent Whittelsey at Harvard.

Serton's papers of the 1920s paved the way for a more comprehensive monograph which was published as a double number of Volume 24 (1933: 1-82) of the *Journal for Economic Geography* under the title *Zuid-Afrika: Een Economisch-geografische Studie* (*South Africa: An Economic Geographical Study*). This was a regional geography against the background of a particular subdiscipline. Serton aimed his work at the informed Dutch public: comparisons with European features were ubiquitous, technical terms were suppressed and economic statistics were artfully weaved into the text. The primary theme was the economic system as developed and maintained by whites. His perspective was Greater South Africa – the subcontinent of permanent white settlement south of the Kunene and Zambezi rivers. The resolution at which he wrote was coarse although less so than Passarge's chapter on *Kolonialwirtschaft*.

Serton commenced with an overview of the subcontinent's natural resource base, then turned to its population and the social conditions, the financial superstructure, the transportation system, industries and commodities and finally, a summary regarding eight economic regions with glimpses of personal observation and a deft sense of place. Pretoria and Johannesburg, Serton (1933: 66) observed, "are two adversaries in the same arena. Pretoria is the seat of government which perceives the City of Gold as its most dangerous opponent ... but has to accept it as indispensable for the national treasury. At the same time the industry finds itself restricted and truncated by the state and its bureaucracy."

An Economic Geographical Study was also the token of Serton's final acceptance of South Africa as home, a white insider's view without harsh value judgments. Notwithstanding its sparkling descriptions the monograph fell into the trap of *geographie humaine*. Problems – droughts, black traditionalism, white poverty – were mentioned but not analysed and the colourful facade obscured the economic challenges of the early 1930s.

Turmoil under the oaks

The Great Depression of the early 1930s hit South Africa with devastating severity. The economy, dependent on overseas trade, came under strain as the Hertzog Government stuck to the gold standard, initiating an outflow of capital and disrupting the prices of export commodities on a flooded world market. Moreover, between 1931 and 1953 the summer rainfall area was scorched by the worst drought of the century, leaving 250 000 poor whites, one in four of the Afrikaners, in its train. Under pressure from Smuts' South African Party the Government left the gold standard in December 1932 and the economy started its slow recovery. With times still serious the two leaders first formed a coalition government and in July 1934 fused their parties, previously two poles apart, into the United Party under Hertzog's leadership. At last Afrikaans and English whites seemed to have found a common cause in rebuilding the shattered economy. However, the right wing of Hertzog's National Party rejected fusion and recast themselves as the Purified National Party under DF Malan. The party was suspicious of the British connections, sternly republican and upholders of an unadulterated Afrikaner culture. By gaining the support of the Afrikaner Broederbond, exclusive, well-organised and eventually country-wide in scope, Malan's followers attained an ideological thrust which the United Party missed. Their power base was in the Cape Province where the Stellenbosch University community was evenly divided between Malanites and Hertzogites (Broodryk 1991).

Stellenbosch, like other South African universities, staggered under the Depression and in the Geography Department student numbers dropped by 30% between 1930 and 1935 but fortunately as the economy prospered from 1936 onwards the university also recovered. With a growing postgraduate presence the Department was now well established and acquired a third and most capable addition in JH Moolman, temporally in 1937 and permanently in 1938.

Serton, rational and pragmatic, welcomed fusion but could not help noticing how the Afrikaners' new confidence was sidelining the Dutch connection. For some Afrikaners Dutch support became superfluous, the Dutch language too stuffy and Dutch forbearance, especially towards the English, too accommodating. These notions were already apparent in 1926 when Serton crossed swords with the Afrikaans author CJ Langenhoven who angrily rejected a call from the South African Academy (*Suid-Afrikaanse Akademie vir Wetenskap en Kuns*) to retain Dutch as a school subject (Kannemeyer 1996). Serton also followed with apprehension the unsavoury heresy trial instigated between 1928 and 1932 by the Stellenbosch Dutch Reformed Church Council against his fellow Utrecht alumnus, the theologian Prof J du Plessis.

In 1937 the Dutch historian Pieter Geyl visited South Africa as a guest of both English and Afrikaans universities and observed how some Afrikaner intellectuals, especially in Pretoria and Bloemfontein, were succumbing to German hypernationalism and racial theory. Afrikaners, he remarked, could not understand that Dutch identity depended on the European balance of power of which Great Britain was the traditional guarantor. Geyl preferred the Western Cape for its spectacular scenery and also for the deeper European roots and open discussion. But in his contacts with Serton and Bouman, the Netherlanders in Stellenbosch, he "discovered considerable tension in this small world" (Van Hees & Willemse 2000: 14). To maintain credibility with Afrikaners, they told Geyl, they could not deviate one iota from the prevailing nationalist dogma on relations with the English and people of colour. Where such a severe ideology might lead was something Serton experienced during a visit to Germany during his sabbatical late in 1938. In the streets of Hannover on 24 October he heard the drumbeat and saw the *Hitlerjugend* marching past: regimented, grim, joyless and so different from South African students enjoying a field excursion. He left Berlin on 5 November, five days before *Kristallnacht* (*Cape Argus* 17 August 1940).

On 4 September 1939 Hertzog's United Party split after a tense parliamentary debate over South Africa's participation in the Western Allies' war against Nazi Germany. The rump of the party, mostly English speakers but with significant Afrikaner support, followed Smuts' leadership into the Second World War as a partner of the Commonwealth. Like Hertzog, most Afrikaners preferred neutrality in what they regarded as a British war and at a meeting in Cape Town in January 1940 Hertzog and Malan decided to merge their following into a Reunited National Party (*Herenigde Nasionale Party*, HNP). Out of power, the Afrikaners split into squabbling factions, their age of discord. While some supported Smuts' United Party, the HNP positioned itself to its right and became the official opposition, republican in sentiment, but still in favour of (white) parliamentary democracy. Further to the right stood the authoritarian Ox Wagon Guard (*Ossewabrandwag*, OB), non-parliamentarian and exalting the *Führer* principle. Beyond the OB were the rancorous right, bizarre splinters like the New Order and Grey Shirts.

Given his background, Serton could hardly be anything else but pro-war. He discovered how painful that was when Nazi Germany invaded the Netherlands on 10 May 1940 and bombed Rotterdam on 14 May. Most Afrikaner intellectuals, some on the Stellenbosch campus, were blatantly insensitive and *Die Burger's* leader on 11 May was an exercise in hypocritical sophistry. "It is the tragedy of the small states of North-western Europe to be trapped among the great powers.... They are the unhappy victims of their geographical position.... Geographically we [South Africa] stand aside. Our interests demand that we should distance ourselves from Europe's wars." In Parliament Malan reiterated this point of view: whatever happened in Europe, South Africa should remain neutral. But for Serton Rotterdam was more than a geographical position. It was the place where he started his career and where relations and friends lived. Fifty years later his colleague Jan Moolman remembered his reaction the day following the bombardment:

> Serton arrived as usual to present his lectures – he never shirked duty. But later during the morning he called us together ... and apologised for not seeing his way open to continue. There was deadly silence while he spoke and he was emotionally drained. Many of his family lived in Rotterdam. I felt profoundly sorry for him. (Personal communication, 15 July 1996).[5]

Serton and his Dutch colleagues, who for years supported Afrikaner political ideals and culture, were painfully disillusioned and turned away in disgust. They ended their association with *Die Burger* and *Die Huisgenoot* and the Sertons withdrew from a Dutch Reformed Church that was riddled with OB sympathizers. Although by then a naturalized South African, he honoured his roots by contributing to Dutch war funds and assisting Dutch citizens marooned in South Africa.

By 1940 Stellenbosch had 10 000 inhabitants, to which the university contributed 70 staff members and 1 200 students. Under the oaks the streets were surfaced and the water furrows, fed from the Eersterivier, were lined with mortared cobbles. Outwardly, the town was as drowsy as ever, but the OB's inflammatory meetings, torchlight parades and rumour mongering kept it in turmoil. Friendships snapped, children accosted each other in the streets and "coloured" houses were plundered in a student riot in July 1940 (Giliomee 2007). Still, an influential Afrikaner minority refused to be swept along, not all of them United Party supporters. In April 1941 Serton was one of 19 lecturers who spoke out against a Transvaal OB scheme to organize a school boycott. "A student who does not know", they asserted (*Cape Argus*, 15 April 1941), "will be a man who does not know and such an ignorant person is... the best subject of a totalitarian state, in which only the leaders are supposed to know."[6]

Few of Serton's students were aware of his discontent. As a teacher he was unemotional, sober and to some extent detached but his lectures were gripping. By 1943 political geography claimed half the third-year course and was presented in bold empirical outlines devoid of the geopolitical chimeras of the day. In the middle of the Second World War, naturally, Serton's perspective was Eurocentric: his lecture was essentially a regional geography of the European family of states and their continental and transoceanic extensions, interpreted in terms of power relationships. Area studies also dominated Serton's course on geographical method for fourth-year research students. In conceptualizing an area study, he compared the German geographers of the 1920s and 1930s in their *Methodenstreit*: first the mavericks Banse and Spethmann, then Passarge's *Landschaftskunde* and finally Hettner and *Länderkunde*, the ultimate model. The French School of regional geography was lauded but not covered in detail.

As research topics most students chose small well-delineated areas – precincts, wards, irrigation districts – and found Serton an accessible and supportive supervisor (Nel 1989). They followed his example, through fieldwork acquired a grass-roots knowledge of their study area, observed visually, conducted interviews and made their own conclusions. No large-scale mapping was expected, no analyses were attempted and the studies lacked historical depth. Yet, the Department's work was recognized internationally when Jan Moolman's paper on irrigation along the Orange River was published in the *Geographical Review* (Volume 36 of 1946).

In 1943, as the tide of the War turned, Smuts and his United Party won what looked like a runaway election victory. But in the same election Malan's HNP rid itself of its radical competitors and, among its lean crop of gains, also acquired the Stellenbosch constituency. While the turmoil subsided and some local discourse revived, Malan's Nationalists retained their ideological rigidity. The innocence of the 1920s was lost for ever. And in the Stellenbosch Department of Bantu Studies colleagues of Serton were developing a megaconcept of their own: they called it "apartheid".

Questioning apartheid

By the 1940s race loomed large in South African political discourse. Stimulated by war, industries mushroomed in the country's four industrial regions. Blacks flocked to the new enterprises, informal settlements burgeoned on the urban peripheries and a revived African National Congress called for residence and union rights. Smuts and his United Party, indulgent during the war, sunk into apathy in the years that followed. The 1948 report of the Fagan Commission (UG 28/48) on black legislation reflected this lack of direction. While dismissing both equality and segregation, the commissioners coined a new term, "parallellism", to acknowledge the permanence of blacks in urban areas. In contrast Malan and his HNP opposition radiated ideological certainty. They replaced "segregation" with "apartheid" and on the eve of the 1948 election a party commission inflated the slogan into a policy: total separation between black and white at local level and a system of independent black "fatherlands" based on enlarged reserves at national level. But for its supporters apartheid had different meanings. Serton's colleagues at the Department of Bantu Studies saw it as an heroic ideal demanding sacrifices of white land and money; for miners on the West Rand apartheid was simply a mechanism to keep blacks in place. Afrikaner solidarity impelled both to vote the HNP into power in 1948.

For geographers the years following the Second World War were a time of growing professionalism. As geography broadened its base toward applied research, young geographers accepted the mesoscale thematic research map as their contribution to the Brave New World of planning. At Stellenbosch Moolman strengthened the practical courses and in his own research processed unpublished population data, based on census wards, into detailed population maps. He left the Department at the end of 1946 and his successor, Andries Nel, applied a similar strategy to the mapping of urban morphology and function. The maps of both researchers emphasized the one overriding feature of South African settlement geography: the contrast in population density between blacks and whites: regionally the overcrowded black reserves opposed to open white farming land and locally black townships as against spacious white urban neighbourhoods.

Serton, belonging to an older generation and tempered by the horrors of two world wars, had a more basic view of race. Arriving in Stellenbosch in July 1920, he accepted the value judgements of his hosts at face value: blacks were thought to be irresponsible in managing their own affairs and those of others, inefficient on the factory floor and trapped in superstition (Serton 1923). However, while visiting Basotholand in 1925 he was impressed by the proud self-reliance of the Basotho and their respect for their nineteenth century King Mosheshwe (Serton 1925). His judgement mellowed. Segregation achieved by separate white and black institutions in different areas, he thought, might be the answer but white and especially Afrikaner thinking on race should become more flexible. He tried to support a change in a brief course on racial geography he presented to third years between 1934 and 1955. Physical anthropologists, he argued, had difficulty in devising a reliable classification of race and found skin colour particularly inadequate. Similarly, "civilization" was not a matter of race but was dependent on history as the head start which the contemporary West had on other cultures. "We should always remember", Serton (Serton & Moolman 1950: 263) remarked drily, "that our North European ancestors were regarded by the Romans as barbarians. Julius Caesar would have smiled at the suggestion that Frisians, Batavians and Britons could attain the same cultural level as Italians."

As statutory apartheid deepened during the 1950s Serton's doubts increased. The National Party, as the HNP rechristened itself, paid lip service to democracy but avoided discourse with its black countrymen. Each measure of Parliament widened the abyss between the races: the Population Registration Act of 1950, the Group Areas Act of 1952 and the constitutional tussle between 1950 and 1955 to remove "coloureds" from the common voters' role, each step rationalized as an unpleasant necessity on the way to the real goal: full rights in the independent black states of the future but no rights in "white" South Africa. In 1955 HF Verwoerd, then Minister of Native Affairs, rejected the key economic proposals of the Tomlinson Commission (UG 61/55) which aimed to provide a decent quality of life for half the country's blacks in the enlarged reserves. The informed public also expected the commissioners to lift the curtain on where and when the reserves would be transformed into "black homelands". Verwoerd clamped down on such speculations and in the process pulled the rug from under the idealists of "Grand Apartheid", people like Jan Moolman, Serton's former colleague and now one of the Tomlinson commissioners, and also his Stellenbosch contemporaries Nic Olivier and Jan Sadie. Debate returned to the Stellenbosch campus and Serton's spirit revived.

The Afrikaner students of the day – optimistic, credulous and self-assured – were different. Political power was now in the hands of the Afrikaner people (*volk*), the students fell into line with their leaders and never doubted that they themselves were the leaders of the future. The select were inducted into covert organizations like the Ruiterwag and Broederbond and proved eager to extinguish whatever deviance which might flare up on an Afrikaner campus. In 1957 Serton was one of seven Stellenbosch lecturers who protested against the Government's intention to establish separate black universities. Their statement was brief and terse: a university should decide on admissions itself. The local National Youth League, the student branch of the National Party, exploded. "For fifty years Stellenbosch was the cornerstone of everything that is Afrikaans. If this (i.e. the admission of students of colour) is what they want, one asks oneself whether their presence here is really desirable" (*Die Pieke*, May 1957). The contrast with the students' welcome message to Serton 37 years earlier and the ignorance of what he did for Afrikaans was stark. Afrikaner nationalism entered its age of arrogance.

Serton still believed that area studies was "the highest form of the geographer's art" and projected his unease about apartheid into two country studies published in the final ten years of his life. In 1950 he accepted an offer from the Dutch publisher Boom to contribute a monograph on South Africa to their very successful series of *Länderkunden*. Now 62 years old, Serton brought his mature judgement and wide knowledge of the country's human geography to bear on the task. The result was *Zuid-Afrika: Land van Goede Hoop (South Africa: Land of Good Hope, 1953)* which went through two editions. As with *An Economic Geography* 17 years before, the book was aimed at the informed Dutch public and again the area under discussion was the subcontinent south of the Zambezi, Greater South Africa in a manner of speaking. But the content was more extensive with chapters on the land, the subcontinent's inhabitants, political organization, sources of prosperity, housing and the relations between different population groups.

Land of Good Hope differed from other regional geographies of the 1950s in its comprehensive review of social and political relations. There was no South African nation, Serton (1955) contended, but a social, economic and political system created by whites over 300 years and perceived as "Western civilization" which they naturally wanted to preserve. The problem was the accommodation of other groups into the system. After discussing Afrikaner, English and African perceptions of themselves and their social milieus he contrasted apartheid and integration in careful, measured language (Serton 1955). Apartheid was an intentional intervention into the system, a procedure which required clear goals. They did not exist. While the extension and integration of black reserves into one black state looked reasonable, "a fully-fledged Bantustan...next to a white South Africa bereft of black labour is an illusion" (Serton 1955: 222). Whites were not mentally and financially prepared to make the sacrifices which such a partition would entail. Integration, on the other hand, was a process, the natural development of a system which already existed. It might be wiser to invest the money and effort which territorial apartheid would demand, to expand the influence of people of colour incrementally and carefully within the system.

Before attending the 1956 conference of the International Geographical Union in Rio de Janeiro, Serton acquired a reading knowledge of Portuguese and afterwards criss-crossed Brazil from Santa Catharina province in the south to Amazonas in the north. As usual, he kept a diary and distilled his impressions in a comparative country study, *Suid-Afrika en Brasilië: 'n Sosiaal-geografiese Vergelyking (South Africa and Brazil: A Social Geographical Comparison*, 1960). The arrangement and content followed *Land of Good Hope* but the book was written in Afrikaans and addressed Afrikaners. "This is a book on South Africa", Serton remarked to friends while working on the manuscript, a look at "... our image in a trans-Atlantic mirror. Everything which holds good here, is bad over there" (Serton Collection). Both countries had a heterogeneous population but the South African economy was better managed and the political culture more stable. In contrast, the mostly white Brazilian elite was open to people of colour who shared their values and gratefully accepted all European immigrants, whether their Portuguese cousins or the culturally more distant Germans and Dutch. Brazilian nationalism and to a large extent Western values permeated all classes of society.

South Africa and Brazil appeared during the same year as the police massacre of pass-book protesters at Sharpeville. The state of emergency which followed demonstrated the African' rejection of the patchwork South Africa envisaged in the Bantu Self Government Act of 1959, a travesty of what the idealists of Grand Apartheid hoped for. Serton's judgement hardened: he referred to "the fictions and prejudices on which the apartheid policy was based", faulted Afrikaners for "a national smugness which regards own opinions as self-evident truths" (Serton 1960: 183) and then posed the question:

> A belief in the almost mythical value of racial purity is a fixed part of South Africa's legacy from its forebears .... Is it really right to make personal happiness wholly subordinate to a theoretically desirable form of social organization? Must we really subject truly human relationships and moral considerations to a rigid ideological policy? (Serton 1960: 202).

Serton's appeal attracted little attention. For *South Africa and Brazil* he prepared a manuscript of 120 000 words; the publisher Oxford University Press wanted 60 000 and eventually the two parties settled for 90 000 (Serton Collection). The dust cover and layout were dull, the text contained no photographs and the six maps were tucked away at the end of the book. While *Land of Good Hope* was well received in the Netherlands – "far ahead of any books on South Africa which appeared in our language" (Van den Broek 1954: 622) – the reviews of *South Africa and Brazil* in the two South African geography journals were trivial and neither mentioned Serton's questioning of apartheid. A South African readership passed Serton by: English geographers did not read Afrikaans and Afrikaners distrusted books from Oxford University Press.

The geographical discourse continues

Serton retired in 1958 and died at his home in Stellenbosch at the age of 75. By that time he commanded wide respect and had collected the honours of a long and productive academic career: in 1928 and 1958 president of the South African Geographical Society, a full member of the South African Academy for Science and Art, board member of the Cape Town-based Van Riebeeck Society, board member of the Dutch Society for Economic and Social Geography and corresponding member of the Royal Dutch Academy of Sciences. But during his last years he became marginalized from both the South African political and geography discourses.

By the 1960s the apartheid state was at its apogee. In the Afrikaner Establishment, dominated by the formidable presence of Verwoerd, political discourse entered its most internalized phase and left Serton the victim of his liberal rationality. His Afrikaner students praised his objectivity and some were even persuaded by his deconstruction of race. But they followed the Establishment which justified apartheid not as a matter of race but of culture. Serton did not advertise his convictions and most students went through his hands without realizing his position on apartheid. Likewise, his Stellenbosch academic colleagues respected his learning and integrity but smiled ruefully at the other-worldly "Hollander" who seemed so out of touch with the South Africa of his day. Meanwhile, in the country of his birth efforts to bring Afrikaners and Dutch closer together foundered. For the Dutch the Afrikaners were no longer the frontiersmen of their "race": they were the outlaws.

Serton was also left behind in the geographical discourse. He still believed that the geographer's task was to provide an accurate image of the Earth's surface: Vidal de la Blache and Hettner were his role models. The young South African geographers of the late 1950s and early 1960s followed the likes of Brian Berry and Peter Haggett and were swept along by the Quantitative Revolution. Some aspired to become demographers, urbanologists, and urban and regional planners. The social engineering of Apartheid South Africa provided chances: new career positions opened in the government's Natural Resources Development Council and its successor, the Directorate of Planning, while young Afrikaners were appointed as lecturers and professors in the geography departments of the new black universities.

Serton produced no master text, pioneered no new themes and made no conceptual breakthroughs. His conception of research lacked a clear methodology and structure, while the value of a book like *Land of Good Hope* as contemporary comment on the South African scene only became apparent 50 years later. But he was an inspiring, thorough and conscientious craftsman whose heritage transcended his years of alienation and ensured continuity at two levels. First and foremost was the Department of Geography which Serton established. From 1920 to 1960 Stellenbosch drew by far the largest number of geography students among South African residential universities and produced 40% of the research degrees. Serton's human geography facilitated the way for younger colleagues who wanted to specialize in population and urban geography, the cutting edges of the Quantitative Revolution. His successor, Andries Nel, was better attuned to shifts in the discipline, accommodated the new subdisciplines in a restructured curriculum and also provided for a modest sliver of physical geography. Geography at Stellenbosch, an American visitor (De Blij 1960: 26) observed, "... has gone ahead vigorously, introducing new courses, employing course assistants on American lines and building up a large and active group of graduate students." The Department clearly had a future.

Less obvious was Serton's introduction of Afrikaans texts into the geographical discourse. Once again Andries Nel was the key link. Because the *South African Geographical Journal* was perceived to be unsympathetic to Afrikaans papers, Nel, with Serton's blessing, initiated a second journal, the *Journal for Geography* (later the *South African Geographer*) which ran for 20 volumes between 1957 and 1993 and carried a 40% load of Afrikaans papers. Afrikaans geographers proved especially creative in urban geography where Van der Merwe's *Die Stad en sy Omgewing (The City and its Environment*, 1983) and Smit & Booysen's (1981) highly topical *Swart Verstedeliking*

(*Black Urbanization*) were undisputed leaders in the field. By 2000, sadly, the Afrikaans braid dwindled into insignificance.

Endnotes

1 Two collections of primary documents were used. The first, and by far the most important, is the Serton Collection which is accommodated in the Manuscript Section of Stellenbosch University's JS Gericke Library. The collection was donated to the Stellenbosch Geography Department by Ms Warda Serton, the youngest of the four Serton children and comprises a selection of Serton's personal correspondence from 1909 to 1963, his diaries of travel, book manuscripts, reviews, class notes and relevant newspaper cuttings. At the time of writing the collection was not indexed. The second collection comprises the minutes of the Council and Senate of Victoria College (1915-1918) and then of Stellenbosch University, all housed in the University Archives.

2 The first time that the title of an Afrikaans or Dutch book or periodical is mentioned, it is rendered in the original together with an English translation. Afterwards only the translation is used. As in Chapter 3 direct quotes are translated into English.

3 Information on the Serton family was provided in conversations the author had on 1-3 June 2009 with Mrs Elizabeth de Wet, daughter of Mrs Gerdien Sem, the second of the Serton children. According to Mrs De Wet the Serton couple, before coming to South Africa in 1920, decided that should Warda not recover from her lung ailment, they would return to the Netherlands.

4 In an interpretation of the intellectual genealogy of Serton's conception of geography, Wesso (1992: 276), who otherwise had great appreciation for his work, asserted that Serton "***had*** to conform, academically", and "***had*** to be responsive to the needs of Afrikaner society." He also claimed that Serton was possibly held captive by the South African academic and socio-economic debate in general and the Stellenbosch one in particular and therefore "***did not feel free*** to develop his own perspective on Geography" (Wesso 1992: 277, author's italics). Wesso provided no evidence for his hunches other than the attention Serton paid to environmental determinists in his fourth-year course and the presence of undergraduate courses on natural regions and race. Because Serton wanted to persuade his Afrikaner audience, both his spoken and written rhetoric were minimalistic, but the author could find no evidence either in the Serton documents or in conversation with his colleagues and former students that Serton was pressured to structure his courses on anything else than his professional judgement. He discussed Ratzel, Passarge and Herbertson because they were important but he was highly critical of most of their ideas.

5 The first reports on the German bombardment of Rotterdam on 14 May 1940 put the fatalities at 30 000 and this must have been the reason for Serton's intense dismay. The true figure was approximately 1 000 but 260 ha of the old city and 25 000 homes were destroyed which left 75 000 persons homeless (Kreikel 1952).

6 The manifesto of Serton and his 18 colleagues on 14 April 1941 was contradicted on 30 April by a counter manifesto of 51 Stellenbosch lecturers who simply said that they disagreed with the first statement. But by not saying what they stood for, the second group neatly sidestepped the difference between the supporters of the HNP and the OB among them. Serton's colleagues DJ Conradie and JH Moolman signed the second manifesto but politics was never discussed in the Department and never clouded the relationship among them (Personal communication: JH Moolman, 15 July 1995).

References

Aay H 1981. Textbook chronicles: Disciplinary history and the growth of geographic knowledge. In Blouet BW (ed) *The origins of academic geography in the United States*, 291-301. Hamden: Archon Books.

Anhaeusser CR 1997. *A century of geological endeavour in Southern Africa 1895-1995.* Johannesburg: Geological Society of South Africa.

ALP 1929?

Baigent E 2004. The geography of biography, the biography of geography: Rewriting the Dictionary of National Biography. *Journal of Historical Geography* 30: 531-551.

Baker AR 1984. *Reflections on the relations of historical geography and the Annales school of history*. Cambridge: Cambridge University Press.

Barnard A 1992. *Hunters and herders of southern Africa: A comparative ethnography of the Khoisan peoples*. Cambridge: Cambridge University Press.

Barnard WS 1992. Öm de Ngami en Makarikari Meren op nieuw te vullen": die Schwarzplan as dwaling. *South African Geographer* 19, 59-75.

Barnard WS 1999. Die weg na 'n omgewingsgeografie. *Acta Academica* 31: 131-161.

Barnard WS 2001. *Conceptions of geography*. Publication No 39/2001. Stellenbosch University: Centre of Geographical Analysis.

Barnard WS 2004. Darwin at the Cape: History of science. *South African Journal of Science* 100: 243-248.

Barnes TJ 2001a. Retheorizing economic geography: from the quantitative revolution to the "cultural turn". *Annals of the Association of American Geographers* 91: 546-565.

Barnes TJ 2001b. 'In the beginning was economic geography'–a science studies approach to disciplinary history. *Progress in Human Geography* 25: 521-544.

Beavon KSO & Rogerson CM 1981. Trekking on: Recent trends in the human geography of southern Africa. *Progress in Human Geography* 5: 159-189.

Beck H 1959. *Gespräche Alexander von Humboldts*. Berlin: Akademie-Verlag.

Beinart W 2003. *The Rise of Conservation in South Africa: Settlers, livestock, and the environment* 1770-1950. Oxford: Oxford University Press.

Beyers CJ 1987. *Dictionary of South Africa Biography.* Pretoria: Human Sciences Research Council.

Botting D 1974. *Alexander von Humboldt: Biographie eines grossen Forschungsreisenden*. München: Prestel.

Boucher M 1974. The University of the Cape of Good Hope and the University of South Africa, 1873-1949. *A study in national and imperial perspective, Archives Yearbook, 35th year, 1972.* Pretoria: Government Printer.

Brown AC 1977. *A History of scientific endeavour in South Africa: A collection of essays published on the occasion of the Centenary of the Royal Society of South Africa.* Cape Town: Royal Society of South Africa.

Buchan J 1903. *The African colony: Studies in the reconstruction.* Edinburgh: W. Blackwood and Sons.

Buttimer A 1993. *Geography and the human spirit.* Baltimore: Johns Hopkins University Press.
Churchill Semple E 1911. *Influences of geographic environment on the basis of Ratzel's system of antropo-geography.* New York, Holt: Archibald Constable.
Crush J 1991. The discourse of progressive human geography. *Progress in Human Geography* 15: 395-414.
Currey RF 1970. *Rhodes University 1904-1970: A chronicle.* Grahamstown: Rhodes University.
Davis WM 1906a. The geographical cycle in an arid climate. *The Geographical Journal* 27: 70-73.
Davis WM 1906a. Observations in South Africa, Bulletin. *Geological Society of America* 17: 377-450.
Davis WM 1906b. The mountains of southernmost Africa. *Bulletin of the American Geographical Society* 38: 593-623.
Driver F 1995. Guest editorial: Geographical traditions: rethinking the history of geography. *Transactions of the Institute of British Geographers* 403-404.
Driver F 1995. Visualizing geography: A journey to the heart of the discipline. *Progress in Human Geography* 19: 123-134.
Driver F 1995. Sub-merged identities: Familiar and unfamiliar histories. *Transactions of the Institute of British Geographers* 20: 410-413.
Driver F 2001. *Geography militant: Cultures of exploration and empire.* Oxford: Blackwell.
Dubow S 1989. *Racial segregation and the origins of apartheid in twentieth century South Africa,* 1919-36. St Antony's: Palgrave Macmillan.
Dubow S 1995. *Scientific racism in modern South Africa.* Cambridge: Cambridge University Press.
Dubow S 2002. Imagining the new South Africa in the era of reconstruction. In Omissi D & Thompson AS (eds) *The impact of the South African war,* 76-95. St Antony's: Palgrave Macmillan.
Dubow S 2006. *A commonwealth of knowledge science, sensibility, and white South Africa, 1820-2000.* Oxford: Oxford University Press.
Du Pisani A 1988. *Beyond the barracks: Reflections on the role of the SADF in the region.* Braamfontein: The South African Institute of International Affairs.
Du Pisani A 1988. *What do we think?: A survey of white opinion on foreign policy issues.* Braamfontein: The South African Institute of International Affairs.
Du Toit AL 1912. *Physical Geography for South African Schools.* Cambridge: The University Press.
Du Toit AL 1924. The contribution of South Africa to the principles of geology. *South African Journal of Science* 21: 52-78.
Du Toit AL 1937. *Our wandering continents: An hypothesis of continental drifting: "Africa Forms the Key".* University of California: Oliver and Boyd.
Eales HV 2007. *Riddles in stone: Controversies, theories and myths about southern Africa's geological past.* Johannesburg: Witwatersrand University Press.

Edel L 1981. The figure under the carpet. In Pachter M (ed) *Telling lives*, 16-34. Philadelphia: University of Pennsylvania Press.
Engelberg E & Schleier H 1990. The history and theory of historical biography-theory appreciation, biographical totality, representational types and forms. *Zeitschrift fur Geschichtswissenschaft* 38: 195-217.
Ferreira OJO 1995. *Adamastor, gees van die Stormkaap, met 'n Afrikaanse verwerking van die Adamastor-gedeelte uit Os Lusíadas van Luis de Camões*. Pretoria: Adamastor.
Foucault M 1970. *The Order of Things: An archaeology of the human sciences*. New York: Routledge.
Fuggle RF 1967. South African geography: An analysis. *The Professional Geographer* 19: 345-351.
Giliomee HB 2003. *The Afrikaners: Biography of a people*. Charlottesville: University of Virginia Press.
Giliomee HB & Mbenga B 2007. *New History of South Africa*. Cape Town: Tafelberg.
Gill D 1905. Astronomy and geology in South Africa. In W Flint & JDF Gilchrist (eds) *Science in South Africa: An Handbook and Review*, 61-78. Cape Town: T Maskew Miller.
Gray JM 1997. Planning and landform: Geomorphological authenticity or incongruity in the countryside? *Area* 29: 312-324.
Groenewald CJ 1984. *Die Institutionalisering Van die Sosiologie in SuidAfrika*. Unpublished doctoral dissertation. Department of Sociology: University of Stellenbosch.
Haggett P 1990. *The geographer's art*. Oxford: Blackwell.
Hammond-Tooke WD 1997. *Imperfect interpreters: South Africa's anthropologists, 1920-1990*. Johannesburg: Witwatersrand University Press.
Hammond-Tooke WD 1997. *Imperfect interpreters: South Africa's anthropologists*. Johannesburg: Witwatersrand University Press.
Hartshorne R 1939. G. The character of systematic geography. *Annals of the Association of American Geographers* 29: 413-436.
Hatch FH & Corstorphine GS 1905. *The geology of South Africa*. London: Macmillan.
Heffernan M 1996. Geography, cartography and military intelligence: The Royal Geographical Society and the First World War. *Transactions of the Institute of British Geographers* 3: 504-533.
Hofmeyr J 1929. The Presidential Address of the South African Association: Africa and Science. *The Times* 27 March 1929.
Hutcheon J 1916. Geography: Its field and its future. *Journal of Geography* 14: 305-306.
Hyam R 1972. *The Failure of South African Expansion, 1908-1948*. New York: Africana Publication Corporation.
Hyam R 1973. Are We Any Nearer an African History of South Africa? *The Historical Journal* 16: 616-626.
Keighren IM 2005. Geosophy, imagination, and terrae incognitae: exploring the intellectual history of John Kirtland Wright. *Journal of Historical Geography* 31: 546-562.

Kellner L 1963. *Alexander von Humboldt.* London: Oxford University Press.
King LC 1962. *The morphology of the Earth.* Edinburgh: Oliver and Boyd.
Kolb P 1719. *Caput Bonæ Spei hodiernum das ist Vollständige Beschreibung des Afrikanischen Vorgebürges der Guten Hofnung.* Germany: Nürnberg
Lambert J 2000. South African British? Or dominion South Africans? The evolution of an identity in the 1910s and 1920s. *South African Historical Journal* 43: 197-222.
Lambert D & Lester A 2006. *Colonial lives across the British Empire: Imperial careering in the long nineteenth century.* Cambridge: Cambridge University Press.
Lester A 2003. Introduction: historical geographies of Southern Africa. *Journal of Southern African Studies* 29: 595-613.
Liebenberg E 2004. *The mapping of South Africa, 1813-1912, in Surveying and Land Tenure at the Cape, 1813-1912. History of Surveying and Land Tenure, Collected Papers, II.* Cape Town: Institute of Professional Land Surveyors and Geomaticians of the Western Cape
Livingstone DN 2000. Putting geography in its place. *Australian Geographical Studies* 38, 1: 1-9.
Lorimer H 2003. Telling small stories: spaces of knowledge and the practice of geography. *Transactions of the Institute of British Geographers* 28: 197-217.
Lückhoff HA 1963. Die Baviaanskloof of Willowmore-seder. *Forestry in South Africa* 3: 114.
Macgregor PA 1955. Development of the theories of the circulation of the atmosphere over South Africa. *South African Geographical Journal* 37: 41-52.
Matless D 1991. Nature, the modern and the mystic: tales from early twentieth century geography. *Transactions of the Institute of British Geographers* 16: 272-286.
Maylam P 1980. *Rhodes, the Tswana, and the British: colonialism, collaboration, and conflict in the Bechuanaland Protectorate, 1885-1899.* Westport: Greenwood Press.
Mentzel OF 1925. *A geographical and topographical description of the Cape of Good Hope.* Cape Town: Van Riebeeck Society.
Molema SM 1920. *The Bantu, past and present: An ethnographical & historical study of the native races of South Africa.* Edinburgh: W. Green & Son.
Muirhead JH 1909. *The central problem of the International Congress on Moral Education.* Hibbert J 6: 346-351.
Myers GA 1998. Intellectual of empire: Eric Dutton and hegemony in British Africa. *Annals of the Association of American Geographers* 88, 1: 1-27.
Obst E & Kayser K 1949. *Die Grosse Randstufe auf der Ostseite Südafrikas und ihr Vorland: ein beitrag zur geschichte der jungen herushebung des subkontinents.* Geographische Gesellschaft zu Hannover.
Pakendorf G 2003. Travellers at the Cape, and what they saw. In von Maltzan C (ed) *Africa and Europe: En/Countering Myths. Essays on Literature and Cultural Politics,* 20-30. Frankfurt am Main: Peter Lang.
Partridge TC & Maud RR 1987. Geomorphic evolution of southern Africa since the Mesozoic. *South African Journal of Geology* 90: 179-208.

Passarge S 1895. *Adamaua: Bericht über die Expediton des Deutschen Kamerun-Komitees in den Jahren 1893-94* (Vol. 2). Berlin: Geographische Verlagshandlung D. Reimer.
Passarge S 1904. *Die Kalahari: Versuch einer physisch-geographischen Darstellung des Sandfelder der südafrikanischen Beckens.* Berlyn: Reimer.
Passarge S 1907. *Die Buschmänner der Kalahari.* Berlyn: Reimer.
Passarge S 1908a. *Südafrika: Eine Landes-, Volks- und Wirtschaftskunde.* Leipzig: Quelle & Meyer.
Passarge S 1908b. Der Tektonik der afrikanischen Küsten. *Petermanns Geographische Mitteilungen* 54: 140-141.
Passarge S 1919. Die Grundlagen der Landschaftskunde... von Siegfried Passarge... L. Friederichsen.
Passarge S 1922. *Beobachtungen über Tier und Mensch.* Leipzig: Quelle & Meyer.
Passarge S 1925. *Grundzüge der gesetzmäßigen Charakterentwicklung der Völker auf religiöser und naturwissenschaftlicher Grundlage und in Abhängigkeit von der Landschaft.* Berlin: Borntraeger.
Passarge S 1951. *Aus achtzig Jahren: eine Selbstbiographie.* Ongepubliseerde manuskrip. Geographisches Institut, Universiteit van Hamburg.
Penck A 1908. Der Drakensberg und der Quathlambabruch. *Königlich Preussische Akademie der Wissenschaften* 11: 230-258.
Penck A 1908. *Das Alter des Menschengeschlechtes.*
Ramutsindela MF 2002. The philosophy of progress and South African geography. *South African Geographical Journal* 84: 4-11.
Raven-Hart R1971. *Cape of Good Hope 1652-1703: The first fifty years of Dutch colonisation as seen by callers.* Cape Town: Balkema.
Rovan J 1995. German pows, a controversial dossier-reply. *Histoire* 193: 92-92.
Raven-Hart R 1976. *Travels at the Cape, 1751-53: An annotated translation of Journal historique du voyage fait au Cap de Bonne-Espérance.* Cape Town: A.A. Balkema for the Friends of the South African Library.
Rogerson CM & Parnell SM 1989. Fostered by the laager: Apartheid human geography in the 1980s. *Area* 21: 13-26.
Rogers AW & Schwarz 1901. *Report on the geology of the Cedarbergen and adjoining country.* Cape Good Hope: Department of Agriculture.
Sandner G 1988. Recent advances in the history of German geography 1918–1945: A progress report for the federal republic of Germany. *Geographische Zeitschrift* 76: 120-133.
Sandner G 1990. *Zusammenhänge zwischen wissenschaftlichem Dissens, politischem Kontext und antisemitischen Tendenzen in der deutschen Geographie 1918–1945: Siegfried Passarge und Alfred Philippson.* In Eckert E (ed), 35-49. Bonn: Philippson-Gedächtnis-Kolloquium.
Sandner G 1992. History of Geographical Thought and Political Context: Reflections on the Commission's Symposia at Bundanoon (1988) and Hamburg (1989). *GeoJournal* 26: 217-218.

Sandner G & Rössler M 1989. *Schriftenverzeichnis von Siegfried Passarge*. Universität, Wirtschaftsgeographische Abteilung.

Sandner G & Rössler M 1994. Geography and empire in Germany, 1871–1945. In Godlewska A & Smith N (eds) *Geography and Empire*, 115-127. Cambridge, MA: Blackwell.

Sandner G 1998. *The response of geography to periods of radical political change: experience from Germany 1918-24 and 1945-50*. Ongepubliseerde lesing. Universiteitskollege, Dublin, 12 November 1998.

Schapera I 1930. *The Khoisan Peoples of South Africa: Bushmen and Hottentots*. Taylor & Francis.

Scheepers GH 1938. *Maisanbau und Klima im sogenannten Maisdreieck der Südafrikanischen Union*.

Schwarz EHL 1905. *Geological Survey of the Coastal Plateau in the Divisions of George, Kynsna, Uniondale, and Humansdorp*.

Schwarz EHL 1906. The Rivers of Cape Colony. *Geographical Journal* 265-279.

Schwarz EHL 1907. *The Rock Channel of the Buffalo River, East London*. Records of the Albany Museum.

Schwarz EHL 1921. *A South-African Geography*. Glasgow: Blackie & Son.

Serton, P. 1923. De haven van Kaapstad. *Tijdschrift voor Economische Geografie* 14, 293-301.

Serton, P. 1925. Basoetoland. *Tijdschrift voor Economische Geografie* 16, 170-181.

Serton. P. 1926. De Zuid-Afrikaansch spoorwegen. *Tijdschrift voor Economische Geografie* 17, 147-148.

Serton, P. 1928a. Snelheid en afstand in die poltieke geografie. *Suid-Afrikaanse Geografiese Tydskrif (South African Geographical Journal)* 11, 5-13.

Serton, P. 1928b. Zuid-Afrika een tweede Sahara? *Tijdschrift voor Economische Geografie* 19, 359-362.

Serton, P. 1929. The desert in human geography. *Suid-Afrikaanse Geograpfiese Tydskrif* (South African Geographical Journal) 12, 26-31.

Serton, P. 1933. Zuid-Afrika, een economische geografische studie. *Tijdschrift voor Economische Geografie* 24, 1-82.

Smit P & Booysen JJ 1981. *Swart Verstedeliking in Suid-Afrika: Proses, patroon en stategie*. Cape Town: Tafelberg.

Smith K 1988. *The changing past: Trends in South African historical writing*. Southern Book Publishers.

Smith N 2003. *American empire Roosevelt's geographer and the prelude to globalization*. Berkeley: University of California Press.

Stigand AG 1923. Ngamiland. *The Geographical Journal* 62: 401-419.

Tamarkin M 1996. *Cecil Rhodes and the Cape Afrikaners: The imperial colossus and the colonial parish pump*. London: Frank Cass & Company.

The Rhodian 1906.

The Stellenbosch Student 1920.

Thomas DSG & Shaw PA 1991. *The Kalahari environment*. Cambridge: Cambridge University Press.

Thomas EM 2006. *The old way: A story of the first people.* New York: Macmillan.
Van der Post C 2005. *Early maps of Ngamiland and the Okavango Delta. Botswana Notes and Records* (Special Edition on Human Interactions and Natural Resource Dynamics in the Okavango Delta and Ngamiland) 37: 196-207.
Warhurst PR 1984. Smuts and Africa: A study in sub-imperialism. *South African Historical Journal* 16: 82-100.
Watt DC 1968. The Decision to Withdraw from the Gulf. *The Political Quarterly* 39: 310-321.
Wellington JH 1929. Some physical influences in the human geography of South Africa. *Suid-Afrikaanse Tydskrif vir Wetenskap* 26: 80-94.
Wellington JH 1932. Land utilization in South Africa. *Geographical Review* 205-224.
Wellington JH 1954-5. *Southern Africa: Physical geography.* Dele I & II. Cambridge: Cambridge University Press.
Wesso HM 1992. *The colonisation of the geographical mind: a critical contextual analysis of the institutionalisation and establishment of geography as an academic discipline in South Africa.* Unpublished doctoral dissertation. Department of Geography, Environmental Studies & Tourism Universtiy of the Western Cape.
Werther R 1935. *Die Kalahari.* Ongepubliseerde tesis. Universiteit van Keulen.
Wilmsen EN 1997. *The Kalahari ethnographies of Siegfried Passarge: Nineteenth century Khoisan and Bantu speaking peoples.* Gaborone: The Botswana Society.

Made in United States
Orlando, FL
05 November 2026

90230144R00057